Rupinder Singh, Ranvijay Kumar, J. Paulo Davim (Eds.)
Piezoelectric Materials for 3D Printing

Advanced Mechanical Engineering

Edited by
J. Paulo Davim

Volume 24

Piezoelectric Materials for 3D Printing

Processing and Applications

Edited by
Rupinder Singh, Ranvijay Kumar and J. Paulo Davim

DE GRUYTER

Editors
Dr Rupinder Singh
National Institute of Technical Teachers Training and Research
Chandigarh 160019, Chandigarh
India
Email: rupindersingh@nitttrchd.ac.in

Dr Ranvijay Kumar
Marwadi University Research Center
Marwadi University
Rajkot 360003, Gujarat
India
Email: ranvijayk12@gmail.com

Prof. Dr J. Paulo Davim
Department of Mechanical Engineering University of Aveiro
Campus Santiago
3810-193 Aveiro
Portugal
Email: pdavim@ua.pt

ISSN 2367-3796
ISBN 978-3-11-914404-9
ISBN 978-3-11-221884-6 (PDF)
ISBN 978-3-11-221913-3 (EPUB)
DOI https://doi.org/10.1515/9783112218846

Library of Congress Control Number: 2026942306

Bibliographic information published by the Deutsche Nationalbibliothek
The Deutsche Nationalbibliothek lists this publication in the Deutsche Nationalbibliografie; detailed bibliographic data are available on the Internet at https://dnb.dnb.de.

De Gruyter and Walter de Gruyter GmbH are part of De Gruyter Brill.
www.degruyterbrill.com

Questions about General Product Safety Regulation:
productsafety@degruyterbrill.com

Cover illustration: Sky_blue/iStock/thinkstock
Printing: CPI books GmbH, Leck

Preface

This book presents the integration of piezoelectric materials with innovative processing routes of 3D printing to deliver well-established information about advancements, fabrication process, characterization, and application-oriented studies. The content of this book is designed for researchers, process engineers, and industry professionals seeking to boost their knowledge of additive manufacturing assisted with piezoelectric materials. The contents of this book are organized to provide basic to advanced knowledge about the synthesis process of piezoelectric materials and their formulation for 3D printing, established hybrid methods for 3D printing of piezoelectric materials, detailed characterizations, and application-oriented case studies. This book focuses on cutting-edge research and industrial applications to eliminate the gap between 3D printing and materials science. This book aims to deliver insightful information for the future of innovative materials and their integration with futuristic additive manufacturing techniques.

 | https://doi.org/10.1515/9783112218846-202

Contents

About the editors

Dr Rupinder Singh is a professor at the Department of Mechanical Engineering, National Institute of Technical Teacher Training and Research, Chandigarh. He has received his PhD in mechanical engineering from Thapar Institute of Engineering and Technology, Patiala. His research areas are additive manufacturing, 4D printing, composite filament processing, rapid tooling, metal casting, and plastic solid waste management. He has authored more than 500 science citation-indexed research papers, 10 books, and more than 100 book chapters, and has presented more than 100 research papers at various national/international conferences. He has been listed among World's Top 2% Scientists, as reported by Stanford University in the years 2022, 2023, 2024, and 2025.

Dr Ranvijay Kumar is an associate professor of research at the Marwadi University Research Center, Marwadi University. He has received his PhD in mechanical engineering from Punjabi University, Patiala. His research areas are additive manufacturing, shape memory polymers, innovative materials, friction-based welding techniques, advanced materials processing, polymer matrix composite preparations, reinforced polymer composites for 3D printing, 4D printing, plastic solid waste management, thermosetting recycling, and destructive testing of materials. He has coauthored more than 100 research papers in science citation index journals, 9 books, and 75 book chapters, and has presented 20 research papers at various national/international conferences. He has been listed among World's Top 2% Scientists, as reported by Stanford University, in the years 2022, 2023, 2024, and 2025.

Dr J. Paulo Davim is a full professor at the University of Aveiro, Portugal. He is also distinguished as an honorary professor in several universities/colleges/institutes in China, India, and Spain. He received his PhD in mechanical engineering, MSc in mechanical engineering (materials and manufacturing processes), and mechanical engineering degree (5 years) from the University of Porto (FEUP), the Aggregate title (Full Habilitation) from the University of Coimbra, and DSc (higher doctorate) from London Metropolitan University. He is fellow (FIET) of IET-London and Eur Ing by Engineers Europe FEANI-Brussels. He is also senior chartered engineer by the Portuguese Institution of Engineers with an MBA and specialist titles in engineering and industrial management as well as in metrology. He has more than 35 years of teaching and research experience in manufacturing, materials, and mechanical and industrial engineering, with a special emphasis in machining, tribology, and surface engineering.

 | https://doi.org/10.1515/9783112218846-204

List of contributors

Abhishek Barwar
Department of Mechanical Engineering
BITS Pilani, Pilani Campus
Pilani, Rajasthan, India

Prateek Kala
Department of Mechanical Engineering
BITS Pilani, Pilani Campus
Pilani, Rajasthan, India

Rupinder Singh
Department of Mechanical Engineering
National Institute of Technical Teachers Training and Research
Chandigarh 160019, Chandigarh, India

Minhaz Husain
Mechanical Engineering Department
Marwadi University Research Center
Marwadi University
Rajkot, Gujarat, India

Gurwinder Singh
Mechanical Engineering Department
National Institute of Technical Teachers Training and Research
Chandigarh 160019, Chandigarh, India

Sourav Kumar
Mechanical Engineering Department
National Institute of Technical Teachers Training and Research
Chandigarh 160019, Chandigarh, India

Ranvijay Kumar
Mechanical Engineering Department
Marwadi University Research Center
Marwadi University
Rajkot, Gujarat, India

Bharat Kalia
Mechanical Engineering Department
National Institute of Technical Teachers Training and Research
Chandigarh 160019, Chandigarh, India and
University Center for Research and Development
Chandigarh University
Mohali, Punjab, India

Ravinder Singh
Mechanical Engineering Department
Chandigarh Engineering College-CGC
Landran, Mohali, Punjab, India

Vinay Kumar
Mechanical Engineering Department
Marwadi University Research Center
Marwadi University
Rajkot, Gujarat, India

Nishant Ranjan
Department of Mechanical Engineering
Faculty of Engineering and Technology
Marwadi University Research Center
Marwadi University
Rajkot 360003, Gujarat, India

 | https://doi.org/10.1515/9783112218846-205

Abhishek Barwar, Prateek Kala, Rupinder Singh*

Chapter 1 Synthesis and processing of piezoelectric-based materials

Abstract: In the recent past, numerous piezoelectric materials were explored for sensing applications, alongside their synthesis and processing techniques. Polyvinylidene fluoride (PVDF), due to its remarkable properties, such as being non-toxic in nature, biocompatible, and resistant to corrosive acids, has been widely used in biomedical applications. Several synthesis techniques, such as electrospinning and polymerisation, are commonly used for preparing PVDF-based fibres, but little is known about the use of biomaterials like hydroxyapatite (HAP) and chitosan (CS) for synthesising and processing through multiple synthesis and processing techniques to ascertain a comparative analysis based on the finally obtained piezoelectric properties. In this chapter, the PVDF has been reinforced with HAP (8%) and CS (2%), and is being synthesised using a stirring method in the dimethylformamide:acetone (60:40) solution to dissolve and further be processed through solution casting and dispenser printing methods to prepare the samples for characterising the piezoelectric behaviour of the PVDF-HAP-CS composite. In parallel, the extrusion-based synthesis method has been explored to facilitate a comparative analysis. Prior to uniform blending, the compound has been processed through the twin- as well as single-screw extruder to attain the uniformity in the diameter, i.e. 1.70 ± 0.05 mm for further processing through the material extrusion process. Several characterisations were performed, including Fourier transform infrared spectroscopy, differential scanning calorimetry, and thermogravimetric analysis, followed by morphological analysis to investigate the composite's piezoelectric behaviour. Based on the comparative analysis, the effective one will be used for piezoelectric sensor applications.

Keywords: Polymerisation, piezoelectric, material extrusion, solution casting, spectroscopy

***Corresponding author: Rupinder Singh,** Department of Mechanical Engineering, National Institute of Technical Teachers Training and Research, Chandigarh 160019, Chandigarh, India, e-mail: rupindersingh@nitttrchd.ac.in
Abhishek Barwar, Prateek Kala, Department of Mechanical Engineering, BITS Pilani, Pilani Campus, Rajasthan, India

 | https://doi.org/10.1515/9783112218846-001

1.1 Introduction

Over the past two decades, piezoelectric materials have become increasingly prominent in various application areas, including medical and healthcare, aerospace, and military, due to their exceptional ability to convert mechanical energy into electrical energy and vice versa. Piezoelectric materials also enhance photochemical and electrochemical properties in response to applied mechanical load [1]. Initially, the quartz crystal was the sole piezoelectric material, mainly used for ultrasonic applications, including submarine detection by echolocation [2]. Over the period, several materials were discovered that also exhibit piezo behaviour, such as Rochelle salt, which is ferroelectric, i.e. it has a domain structure, spontaneous polarisation, and can exhibit more than one phase that can be changed with the application of an electric field. But the drawback of Rochelle salt is that its ferroelectric properties may diminish even with a small variation in the composition [3]. To overcome these challenges, other materials with excellent ferroelectric properties were discovered over time, like barium titanate ($BaTiO_3$), lead titanate ($PbTiO_3$), and lead zirconium titanate (PZT), and found useful in various fields such as ultrasound imaging, signal processing, smart structures, sensors, actuators, and generators [4]. Piezoelectric materials are categorised into two groups: natural and man-made. Materials like quartz, topaz, wood, silk, and Rochelle salt are natural piezoelectric materials, whereas ceramics such as $BaTiO_3$, PZT, and potassium niobate ($KNbO_3$) [5] and polymers such as polyvinylidene fluoride (PVDF) and polydimethylsiloxane (PDMS) are man-made piezoelectric materials [6]. In recent years, piezoelectric materials have been found suitable for energy-harvesting applications, specifically for harvesting green energy from various sources in our surroundings [7]. Researchers have suggested materials such as PZT, BaTiO3, and PVDF for piezoelectric nanogenerator applications due to their greater potential for harvesting mechanical energy for applications such as sensors, wireless communication, and battery charging [8, 9]. PVDF, due to its excellent stability, high piezoelectric coefficient, high flexibility, and other properties, is particularly well suited for nanogenerator applications [8, 10]. Recently, many preparation methods have been explored to fabricate PVDF films for sensing or supercapacitor applications. Among the various synthesis methods, electrospinning is considered the most significant due to its low cost, simple operation, and low applied voltage; therefore, it is desirable for the PVDF phase during any loading application [11]. This technique is suitable for attaining high-performance films by subjecting the material to a high electric field and collecting the fibrous film with a variable diameter ranging from microns to nanometres [12]. Alongside electrospinning, solvent casting, tape casting, phase separation, and hot pressing, other synthesis techniques are intensively explored to fabricate thin PVDF films [13, 14]. Hot press is another technique used for the preparation of the PVDF and its co-polymers films. Using the hot-press method ensures complete solvent evaporation, even without dissolving the sample in the organic solvent [15, 16]. The critical parameters for the hot-press method are temperature, pressure, and operating time, depending on the experimental conditions [17]. It has certain benefits over other approaches, such as eliminating the voids in the matrix caused by rolling induced by high temperature and pressure, which has been shown to improve mechanical properties [18]. Previous studies reported the solvent evaporation method as one of the simplest methods for developing thin films of PVDF [19]. It was found that the crystalline structure of PVDF depends on the rate of solvent evaporation [20]. Also, it was observed that the high-porosity cast PVDF exhibits improved ionic conductivity but has a lower β-phase content due to its lower piezoelectric coefficient [21, 22]. Irregular shapes, complexity of the process, additional polling, and mechanical stretching of the obtained films are required to obtain the β-phase with a higher piezoelectric response [23]. Material extrusion (MEX)-based additive manufacturing can also induce β-phase content in PVDF under stress, but it is insufficient to achieve a high β-phase content (>50%) [24]. Varying additive proportions in PVDF or optimising printing parameters are found to be effective for enhancing the β-phase content, but post-processing polling is also needed to align the dipoles for achieving macroscopic piezoelectricity [25]. MEX-based fabrication offers several advantages, including ease of fabricating complex or intricate structures, reduced material waste, and a multilayered design that enhances the d_{33} of the fabricated device [26]. Melt-processable fluoropolymers are found suitable for high-performance engineering applications, and PVDF is the only fluoropolymer that is produced from a single repeating structure (VF_2) [27]. For many applications, PVDF is processed by injection moulding and extrusion, where the polymer specifications are designed, so that molecular weight controls the melt viscosity. Good chemical and oxidative stability, along with a combination of properties and processability, make it one of the most widely used fluoropolymers [28]. It covers a wide range of applications, including nuclear power engineering, the aerospace industry [29], electronics and radio engineering [30], cable insulation coating, solar collectors, and printed circuit boards [31]. Recently, radio frequency devices, such as microstrip patch antennas (MPAs), have been reported to use PVDF as a dielectric material for wearable or implantable applications due to its flexibility and piezoelectric behaviour [32]. In recent years, natural biopolymers such as hydroxyapatite (HAP) and chitosan (CS) have been used for scaffolding and orthopaedic applications due to their excellent biodegradability and biocompatibility [33, 34]. Recent advancements in technology have led to the development of new materials with improved chemical, mechanical, and physical properties for the regeneration or reconstruction of tissues inside or outside the body [35]. CS (a partially deacetylated form of chitin) added to a polymer enhances bioactivity and improves cell adhesion and differentiation at the host site [36]. HAP (derived from eggshell) made scaffolds are biocompatible and bioactive, have a porous structure, and are therefore suitable for bone regeneration applications [37]. Reinforcing HAP and CS into PVDF at 8% and 2%, respectively, enhances the mechanical strength of the end component for high-flexibility and loading applications [38]. MPAs have recently gained significant interest due to their numerous advantages, such as ease of fabrication, low cost, simple design, and high-quality communication [39]. PVDF, due to its piezoelectric nature and flexible properties, may be used for wearable or implantable antenna applications. Recently, several synthesis and processing techniques have been used to process PVDF and its composites for piezoelectric and energy-harvesting applications, but hitherto little has been reported on its use for the fabrication of implantable patch antennas through different approaches, specifically solution casting and MEX, and their comparative analysis.

1.2 Research gap and bibliometric analysis

Further, to enrich the literature review, a detailed bibliometric analysis has been conducted using data extracted from the Scopus and Web of Science (WOS) databases over the past two decades. Figure 1.1(a) shows the trend in research on the processing and synthesis of PVDF and other polymers over the years, as measured by the number of documents reported per year. The data show that the evolution of techniques, such as electrospinning and MEX, has a significant impact on the rate of reported documents; e.g. in 2013, 17 documents were reported on the synthesis and processing of piezoelectric materials. The trend remained steady until 2022; thereafter, it increased steadily and peaked in 2025, with 26 documents reported in a single year. Figure 1.1(b) highlights the nature of the documents reported over the years: 52.9% are research articles, followed by 13% conference papers, 11.2% review papers, and 9.4% book chapters. To observe the country-wise contributions in the respective fields, it was found that the USA leads with 40 documents, followed by India and China (Figure 1.1 (c)). Observing the most influential areas that contributed the most to the exploration of the piezomaterials. Figure 1.1(d) shows that material characterisation techniques, such as Fourier transform infrared spectroscopy (FTIR), X-ray diffraction (XRD), and the morphological analyses, proved helpful in the identification of piezo properties as well as the synthesis of the material followed by the engineering applications (i.e. strain sensors, implantable sensors, and energy harvesting) of the materials processed through various techniques like solution casting, hot press, electrospinning, and extrusion-based methods. Similarly, Figure 1.1(e) represents the contributions of the top 10 authors working in these areas based on the number of documents reported.

Further, the network diagrams were prepared based on WOS data for 196 articles, from which a total of 12,810 terms were selected for the investigation. During the analysis, the minimum term occurrence was set to 5, and based on this, 618 key terms met the threshold, of which 60% were the most relevant terms used to prepare the bibliometric diagram (Figure 1.2(a)). The network diagram consists of four clusters, differentiated by different colour codes, each highlighting specific areas, like the green cluster represents the various material characterisation techniques, i.e. XRD, Field Emission Scanning Electron Microscopy (FESEM), Transmission Electron Microscopy (TEM), and Transmission Electron Microscopy (EDX), the blue cluster shares the infor-

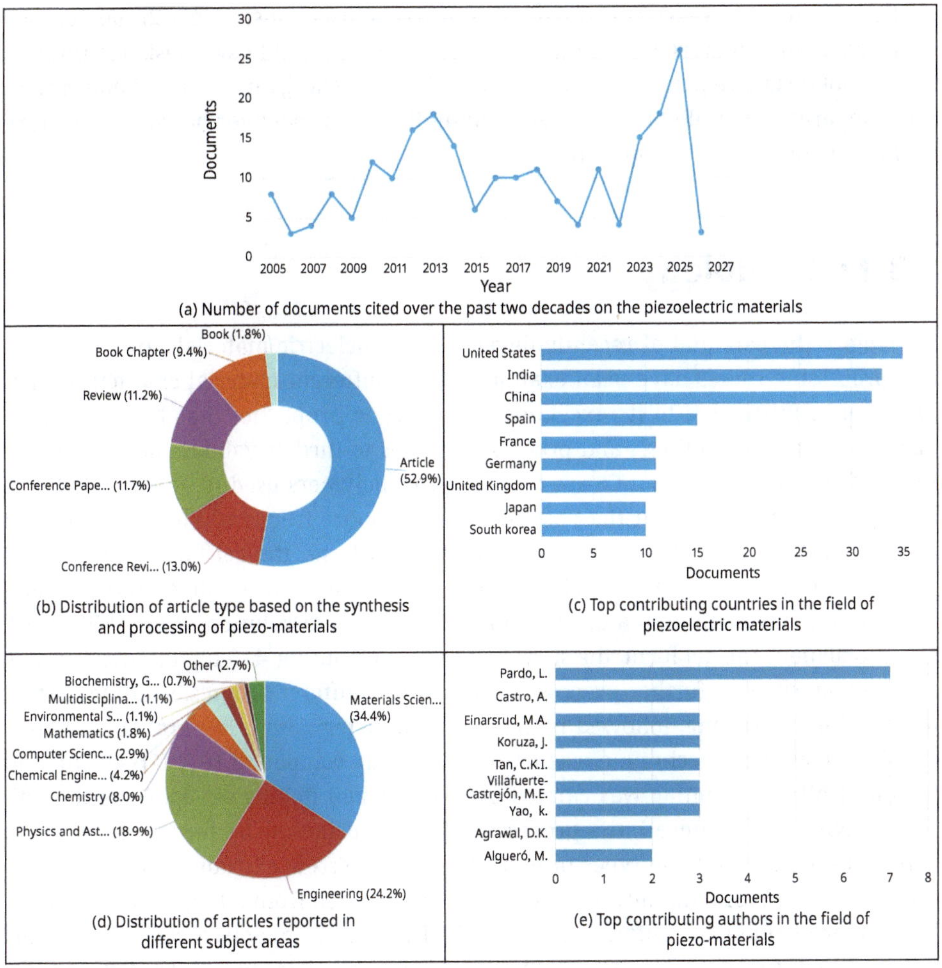

Figure 1.1: Bibliometric analysis of the PVDF, its synthesis, and processing routes based on the Scopus database for the past two decades.

mation of the types of materials and their processing techniques, whereas the red cluster gives information on the various applications of the piezoelectric materials, i.e. nanogenerators, energy harvesting, and sensing. Upon highlighting the node "development" in the red cluster in another network diagram (Figure 1.2(b)), it has connections with several nodes representing the areas in which most of the research has been carried out over the years, i.e. synthesis techniques like piezocatalysis and photocatalysis used to prepare the nanocomposites further characterised with the spectroscopy and XRD to identify the piezoelectric nature but limited has been identified on the application of extrusion-based techniques to prepare and process the piezomaterials. Similarly, another node, i.e. "ceramic", has been highlighted to explore its connections with other key terms (Figure 1.2(c)). Previously, most research has focused on piezoceramic materials, such as PZT, with many available synthesis techniques, such as solid-state reaction and the sol-gel method, for energy-harvesting applications, but comparatively, little has been reported on the use of polymer-based piezomaterials for advanced sensing applications.

1.3 Methodology

To explore the potential of recently developed piezoelectric materials, two synthesis and processing routes have been examined using different material characterisation techniques. PVDF, due to its excellent piezoelectric properties, has been analysed based on different synthesis and processing routes to further validate its use for biosensor fabrication. HAP and CS are two bioactive polymers used to enhance cell proliferation and bioactivity at the host site, and are more frequently used for implant and scaffolding applications. Primarily, for the synthesis, the gel formation method involves dissolving PVDF-HAP-CS (20 g) in a solvent mixture of dimethylformamide (DMF) and acetone in a 60:40 ratio. For complete dissolution of the PVDF-HAP-CS or to achieve the gel form, the stirring was carried out for 4–6 h [40]. After synthesis, the material was further processed through two different routes: solution casting and dispenser printing, followed by material characterisation. In parallel, an extrusion-based synthesis technique was used to form the compound (PVDF-HAP-CS), and for this, uniform blending was carried out using a ball milling set-up. Once the uniform mixing was achieved, the processing, i.e. extrusion of the feedstock filament, was carried out using two types of extruders: single-screw extruder (SSE) and twin-screw extruder (TSE). Initially, the strands were extruded from a twin-screw extruder for further blending or uniform mixing of the HAP and CS nanoparticles in the PVDF, followed by the single-screw extrusion to maintain a diameter to be 1.70 ± 0.05 mm for smooth operation of fused filament fabrication (FFF). The samples were prepared for further characterisation of the prepared PVDF-based composite. Based on the characterisation results, a better synthesis and processing route (in terms of piezoelectric properties) has been adopted for sensor fabrication. Figure 1.3 highlights the detailed methodology adopted in this study.

1.4 Experimentation

1.4.1 Synthesis of PVDF-HAP-CS

The synthesis of PVDF-HAP-CS has been carried out using two approaches: gel preparation and MEX. The gel preparation method, or the polymer dissolution method, in-

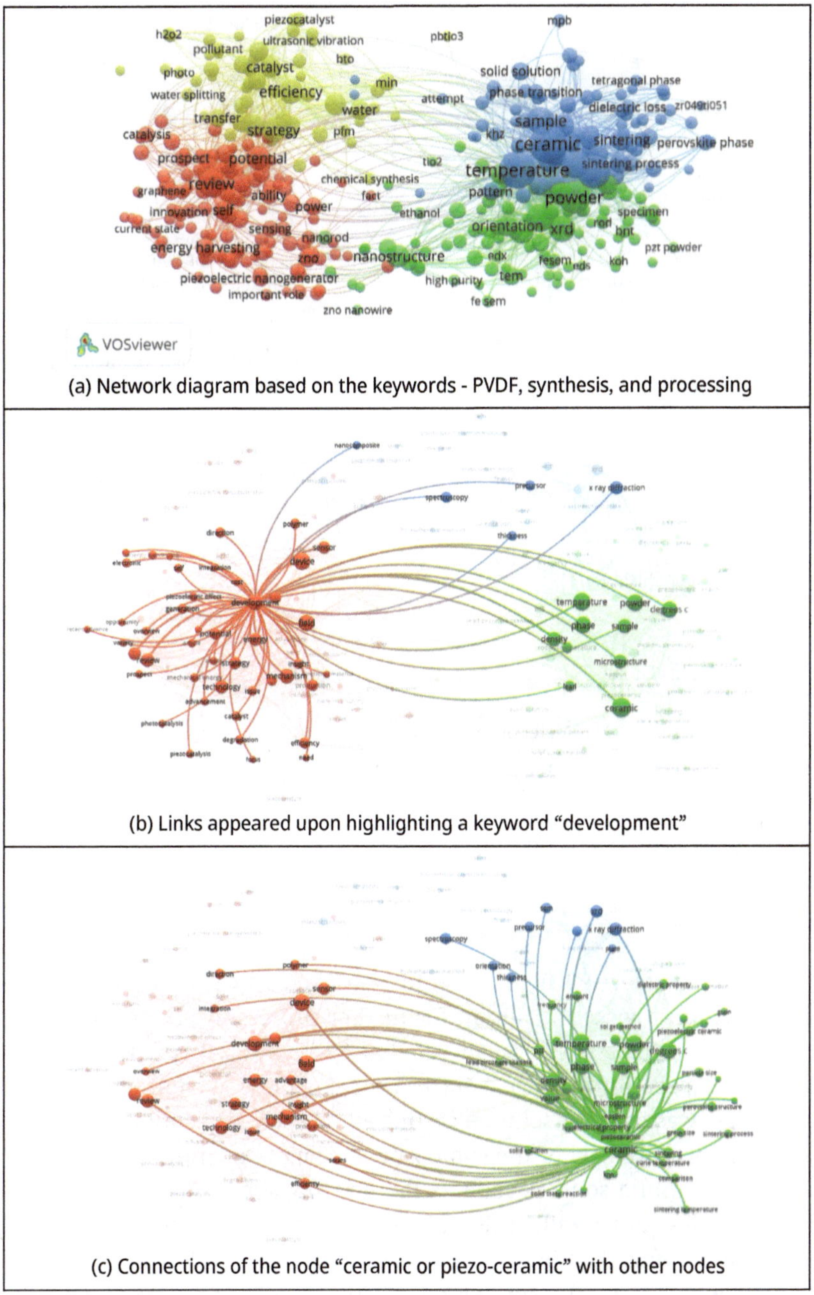

(a) Network diagram based on the keywords - PVDF, synthesis, and processing

(b) Links appeared upon highlighting a keyword “development”

(c) Connections of the node “ceramic or piezo-ceramic” with other nodes

Figure 1.2: Co-occurrence analysis of the various terms searched on the Web of Science database.

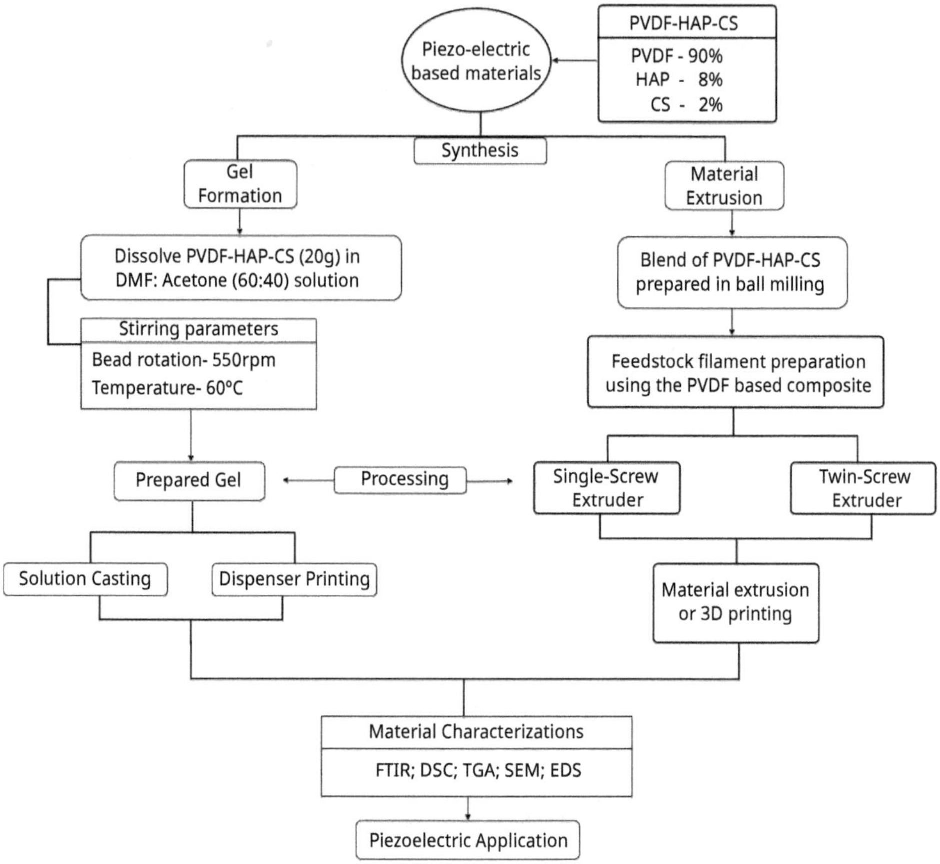

Figure 1.3: Methodology adopted for this study.

volves dissolving the polymer (PVDF) and the fillers (HAP and CS) in a DMF:acetone (60:40) solvent to obtain a viscous solution for further processing. The solution has been prepared using a magnetic stirrer operated at 60 °C and 620 rpm for 4–8 h to achieve complete evaporation (Figure 1.4(a)). After complete solvent evaporation (Figure 1.4(b)), the gel was used for solution casting and dispenser printing. The solidification time for the prepared gel of the composite at room temperature is very low, and therefore it turns into a solid as soon as the printing starts (Figure 1.4(c)). For MEX, the PVDF granules, along with HAP and CS nanoparticles (Figure 1.4(d)), were poured into the ball-milling chamber and processed for 3–4 h at 80 rpm [41]. Once uniform mixing was achieved, the compound (PVDF-HAP-CS) was used to prepare feedstock filaments primarily from the TSE to obtain uniform blending of the HAP and CS nanoparticles inside the PVDF (Figure 1.4(e)). The operating parameters of the extruder are: screw speed 100 rpm and temperature 210 °C, whereas the filament diameter remained uneven throughout its length. To further achieve uniformity in the diameter, the prepared strands were mechanically shredded (Figure 1.4(f)) into small units and then poured into the hopper of the SSE. The operating parameters are: screw speed, 6 rpm; temperature, 210 °C; and filament diameter, 1.70 ± 0.05 mm (Figure 1.4(g)). The prepared filaments were then processed through the MEX printing for sample preparation.

1.4.2 Processing of PVDF-HAP-CS

The processing of the synthesised PVDF-based composite has been further carried out using multiple approaches, namely solution casting, dispenser printing, and MEX printing. Primarily, the composite gel was poured into the desired shape cavity (i.e. to create the sample), and the excess thickness was controlled using the doctor-blading technique (Figure 1.5(a)). To remove the solvent from the cast object, it was placed in a vacuum-controlled temperature oven for a few hours to achieve uniform entanglement of the polymer chains and a smoother substrate. The curing and drying process not only removes the solvent but also allows the polymer chains to reorient, thereby reducing internal stresses. Once drying and curing were complete, the substrate was peeled from the acrylic glass plate and used for further characterisations. For performing the MEX printing, the selective parameters like the printing speed, bed temperature, nozzle temperature, and layer height were optimised to achieve sound printing. Bed adhesion and continuous extrusion were the major issues encountered during the printing of PVDF and its composites (Figure 1.5(b)). To tackle these issues, the printing speed was reduced to 50%, while the flow rate was increased to 150%; other parameters, such as bed and nozzle temperatures, were set to 90 and 230 °C, respectively. The nozzle diameter was 0.6 mm, the layer height was 0.16 mm, the infill density was 100%, and the infill pattern was rectilinear. The samples have been prepared for various characterisations, and a comparative analysis has been conducted.

1.4.3 Different characterisations on PVDF-HAP-CS

Based on the synthesis and different processing routes, the prepared samples were tested using various techniques, namely differential scanning calorimetry (DSC), thermogravimetric analysis (TGA), morphological, and FTIR analysis (Figure 1.6(a)–(d)). To ensure the presence of different functional groups and the chemical structure of the prepared composite, FTIR with attenuated total reflectance assembly (Make: Bruker) has been performed within the 400–4,000 cm^{-1} spectral range. The transmittance % has been observed at different wavenumbers, which can help identify the effects of synthesising and processing the composite via different routes. Further, to ensure the thermal stability of the prepared composite, its thermal transitions were examined using a DSC (Mettler Toledo, Make: TA Instruments) with three consecutive cooling

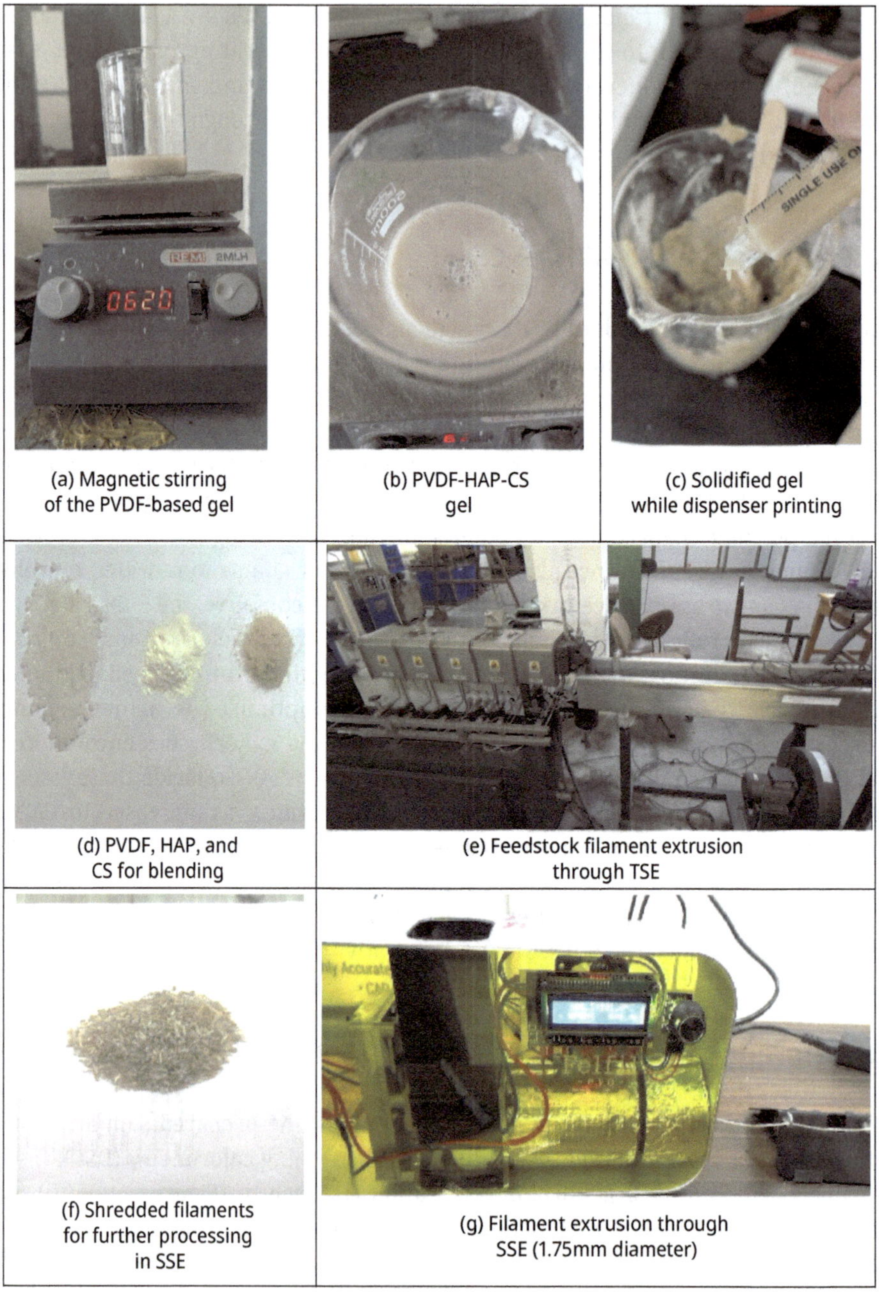

Figure 1.4: Steps involved in the synthesis of PVDF-HAP-CS composite through solution preparation and wire extrusion methods.

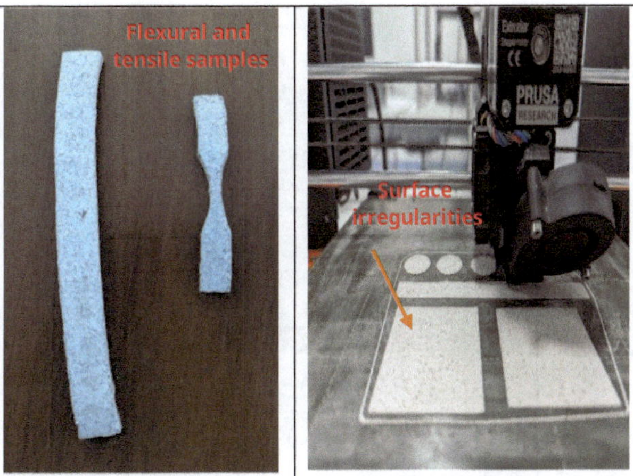

(a) Solution-cast samples

(b) MEX printing of the composite

Figure 1.5: Processing methods and the samples prepared with two approaches.

and heating cycles, over a temperature range of 35–300 °C at a heating rate of 10 °C/min. The samples from both synthesis routes, i.e. gel and MEX, are weighed at 7.6 and 8.6 mg, respectively, prior to performing the experiment. Apart from this, the weight loss or the thermal degradation of the composite with time has been measured through a TGA set-up (Mettler Toledo, Make: TA Instruments) over a range of temperatures, i.e. 35–700 °C at a heating rate of 10 °C/min. The analysis has been carried out on the samples, weighing 8.57 and 8.06 mg for the gel and MEX approaches, respectively, under a nitrogen environment (supplied at 20 ml/min). Further, to understand the dispersion of HAP and CS nanoparticles in PVDF, SEM analysis (Make: JEOL) was performed, and the elemental composition was verified using an EDS system (AZtecOne, Make: Oxford Instruments). Several images have been captured at different magnification scales.

1.5 Results and discussion

1.5.1 FTIR spectroscopy

Figure 1.7 shows the transmittance spectra of the composites prepared using the gel and MEX approaches. The transmittance peaks observed in both cases are nearly identical, with only a small variation, suggesting the possible presence of functional groups. FTIR has proven effective for detecting different PVDF polymorphs (i.e. α, β, and γ) and for assessing the effects of various processing parameters on the properties of the semi-crystalline polymer. PVDF composite, processed with an extrusion-based method,

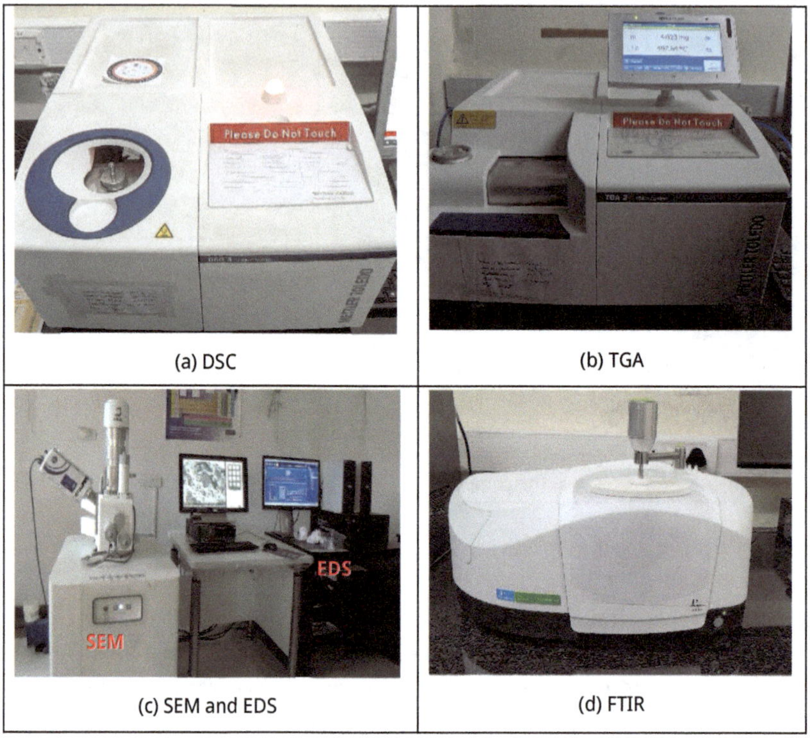

Figure 1.6: Experimental set-ups used for various characterisations performed on the prepared composite.

showcased the transmittance peaks at 532.58, 615.45, 763.38, and 873.86 cm^{-1}, showing the presence of the α-phase, whereas a peak around 1,403.58 shows the presence of the β- and γ-phases [42]; therefore, the electroactive nature of the PVDF. The molecular chain conformation was done based on the various absorption bands, i.e. a peak at 2,923.38 confirmed the presence of hydrogen bonds, peaks in 800–600 (615.45 and 763.38), 1,200–1,000 (1,178.93 and 1,061.69), and 1,600–1,300 cm^{-1} (1,403.58) highlight the presence of the hydroxyl (OH) or amino groups [43], which further confirms the traces of the CS for enhancing the bioactivity, and anti-inflammation reactions and could be beneficial for the fabrication of the implantable sensors. Further, a sharp peak at 1,668.13 cm^{-1} indicates the presence of carbonyl compounds (C=C) [43], providing evidence of mechanical stretching of the compound during the extrusion process. A small peak is detected between 873.76 and 1,061.69 (i.e. around 950 cm^{-1}), indicating the presence of vinyl terminals ($-CH=CH_2$) [43]. Therefore, the overall structure of the PVDF-HAP-CS has been counter-verified by identifying the various functional groups through infrared spectroscopy, which shows vibrational bands in the molecular structure. While looking over the IR spectra of the solution-cast sample, sharp, intense peaks were observed as compared to the MEX-processed sample with a minor shift in the wavelength except for a few. Transmittance peaks at 480.41 and 837.03 cm^{-1} highlights the presence of the electroactive γ-phase [42], and therefore dominating the α-phase. Other peaks are almost overlapping the behaviour of the composite in case of MEX except for the intensity of transmittance. The observations from the FTIR spectroscopy suggest that the solution cast method may be more suitable for the piezoelectric applications like wearable or implantable biosensors.

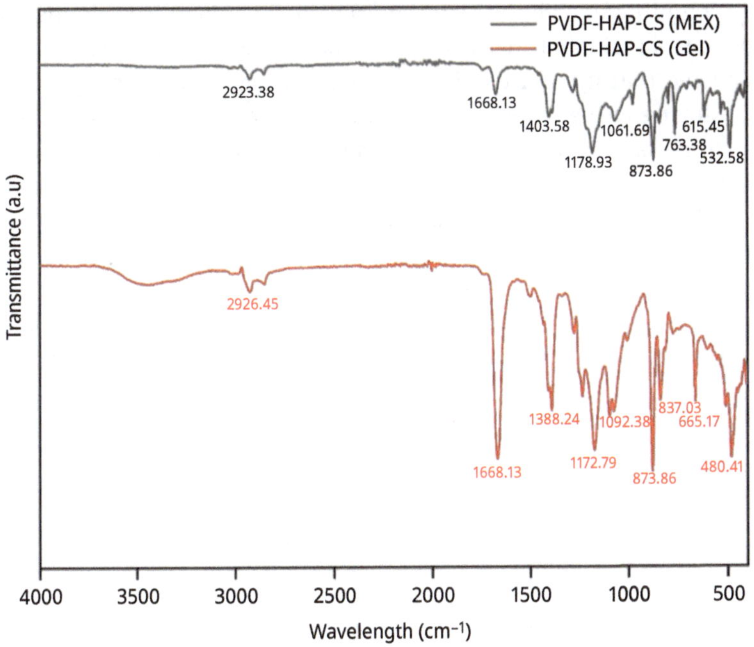

Figure 1.7: FTIR (transmittance spectrum) of the samples prepared with MEX and solution casting.

1.5.2 Thermal characteristics

To analyse the thermal properties, the samples have been tested using DSC and TGA. During the DSC analysis, the sample underwent three consecutive heating and cooling cycles to ensure repeatability and to remove the residuals from previous experiments performed in the set-up. The graphs were differentiated by colour codes: red for gel-processed samples and black for MEX-processed samples (Figure 1.8). Upon observing the behaviour of the MEX sample, it was found that it requires 33.5 J/g to initiate melting, with an onset temperature of 162.88 °C and an endset temperature of 172.03 °C. Observations from the exothermic reaction indicate that 42.61 J/g of heat was released by the sample as it reached room temperature from an onset temperature of 142.19 °C to an endset value of 135.19 °C. The initial cycle was intended primarily to clear the previous residual. Moving on to the second endothermic cycle, it was observed that 43.67 J/g of heat was required to melt the sample, with an onset value of 157.94 °C and an endset value of 171.88 °C. The exothermic reaction shows that 42.76 J/g of heat was released as the sample approached room temperature from an onset temperature of 142.20 °C to an end value of 135.62 °C. Further, the third heating and cooling cycle almost replicated the results obtained in the second cycle. Looking over the results from the gel-processed samples, it was observed that in the primary endothermic cycle, 264.95 J/g was required to melt the material from 46.77 to 116.97 °C. This absurd behaviour was due to residual impurities from previous experiments. The exothermic reaction results showed that it released 31.42 J/g over the temperature range from 141.43 to 133.65 °C. Moving to the second cycle, it was observed that 32.35 J/g of heat was required to melt the sample from 157.60 °C to an endset temperature of 170.66 °C, whereas the exothermic reaction indicated that it released 31.09 J/g from an onset temperature of 141.45 °C to an endset temperature of 134.27 °C. Like the MEX-processed sample, the gel-based sample also shows behaviour quite like that in the third thermal cycle. Overall, the DSC analysis indicates that the composite samples behaved similarly under both processing methods, highlighting the material's suitability for high-temperature applications and its ability to undergo other high-temperature processing.

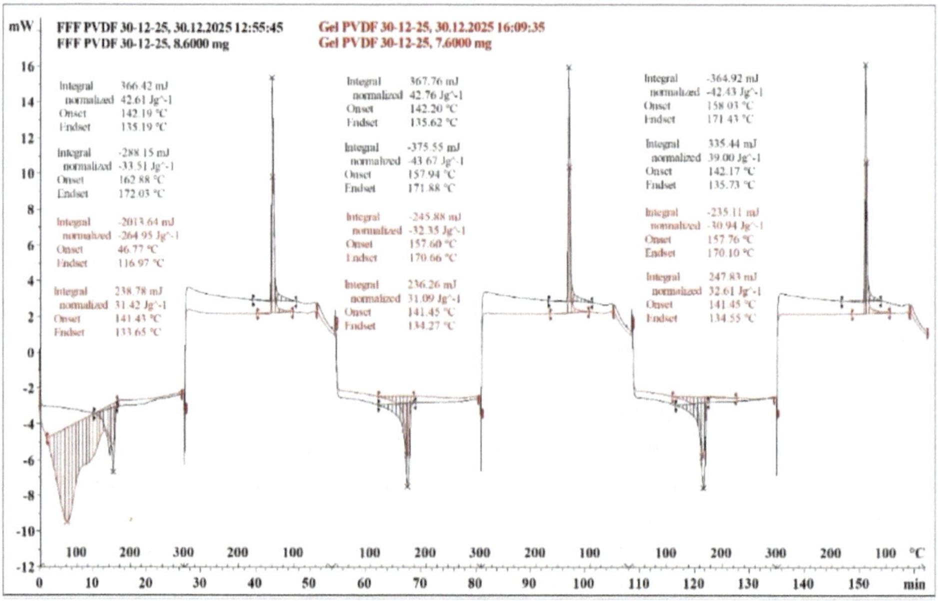

Figure 1.8: DSC analysis of the prepared composite with multiple approaches.

The weight loss or degradation of the prepared composite with increasing temperature has been analysed by TGA (Figure 1.9). The MEX-processed sample weighing 8.09 mg has been tested over a temperature range from 35 to 700 °C. It was observed that the sample exhibited one decomposition stage, from the onset temperature of 350.28 °C to the endset temperature of 600.30 °C, which may have happened due to the loss of dopant and the bonded water, and after that, the polymer chain structure began to degrade, leaving the material to completely decompose around 680 °C. In the gel-processed sample, the composite exhibited two decomposition stages, primarily due to moisture and solvent residues, beginning at 50 °C and extending to 150 °C, then shown almost similar behaviour was shown by the MEX-processed sample. The reinforcement of HAP and CS nanoparticles enhanced the composite's thermal stability.

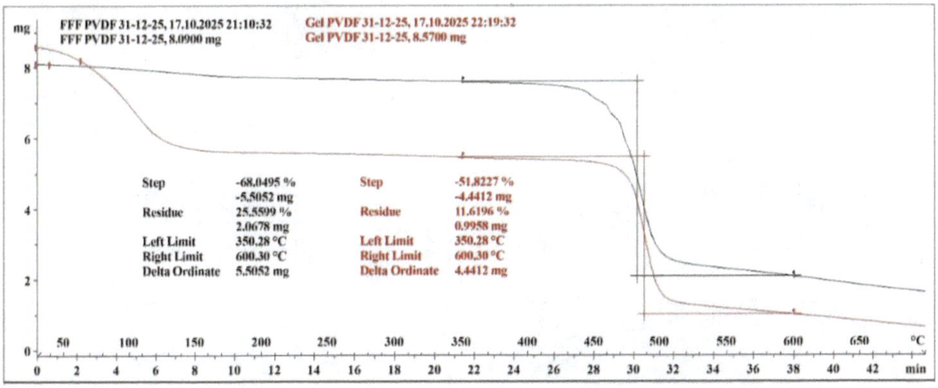

Figure 1.9: TGA of the prepared samples.

1.5.3 Morphological analysis

Upon examining the morphology of the samples processed through two approaches, it was observed that the MEX sample has uneven surfaces comparatively (Figure 1.10(a) and (b)), also it exhibits the absence of the CS nanoparticles over the observed area, counter-verified with the EDS data (i.e. elements like Ca and N are not detectable) (Figure 1.10(c)), therefore lacking the capability of enhanced bioactivity. The reinforcement of HAP and CS nanoparticles is desirable for implantable or wearable applications due to their attractive mechanical and biocompatible properties, but processing the prepared composite via the MEX route is difficult, as the nanoparticles may clog the nozzle, leading to interruptions in continuous printing and ultimately resulting in surface irregularities. The measured average surface roughness ($R_a = 14.10 \times 10^{-3}$ μm) for the MEX process (Figure 1.10(e)) also highlights the challenges of depositing conductive ink to form circuits. The gel-cast sample exhibits the agglomeration of the HAP and CS nanoparticles over specific portions of the sample surface. Looking over the elemental spectrum, the composite was observed to exhibit a higher O wt% (6.92) (Figure 1.10(d)), which may be responsible for the 2-stage decomposition observed during the TGA. Also, the solution-cast sample exhibited lower surface roughness (R_a=6.05 × 10^{-3} μm) (Figure 1.10(f)), which is desirable for sensor fabrication.

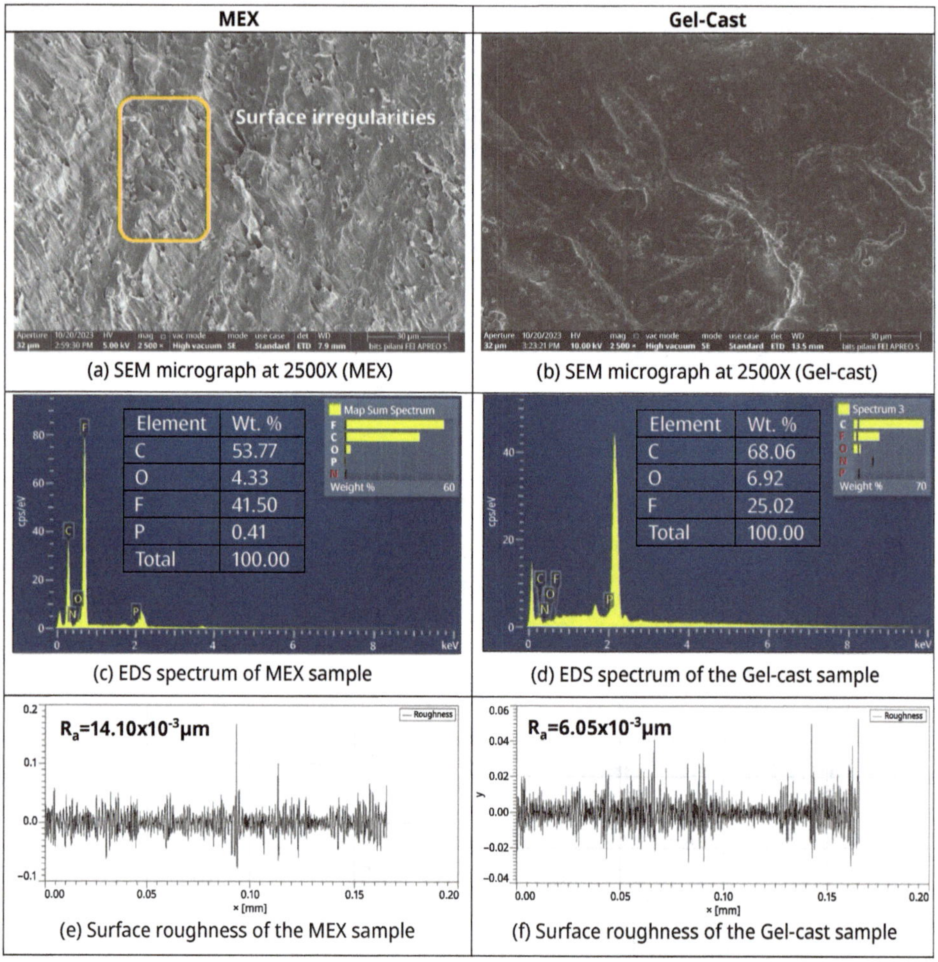

(a) SEM micrograph at 2500X (MEX)
(b) SEM micrograph at 2500X (Gel-cast)
(c) EDS spectrum of MEX sample
(d) EDS spectrum of the Gel-cast sample
(e) Surface roughness of the MEX sample
(f) Surface roughness of the Gel-cast sample

Figure 1.10: Morphological identifications of the prepared samples with MEX and gel-cast methods.

1.6 Summary

Investigations into the synthesis and processing of piezoelectric materials provide insights into the challenges encountered when handling them across various approaches. The infrared spectroscopy results highlights that the composite processed with MEX showcased peak around 1,403.58 that highlight the presence of the β- and γ-phases; therefore, the electroactive nature of the PVDF other than the transmittance peaks at 532.58, 615.45, 763.38, and 873.86 cm^{-1}, showing the presence of the α-phase, whereas, PVDF-HAP-CS processed with solution casting intense peaks were observed as compared to the MEX-processed sample at 480.41 and 837.03 cm^{-1} highlights the presence of the electroactive γ-phase, and therefore dominating the α-phase. The thermal analyses suggest that the composite behaved quite similarly under both methods over three consecutive endothermic and exothermic heating cycles, whereas the TGA highlighted the presence of moisture and the solvent in the solution-cast sample, which may be responsible for its two-stage decomposition before the complete degradation of the material. The morphological analysis of the samples indicated that the reinforcement of HAP and CS nanoparticles in the PVDF matrix leads to nozzle clogging, thereby interrupting continuous material deposition and causing surface irregularities. The MEX possesses a higher R_a (14.10×10^{-3} μm) than the solution-cast sample (6.05×10^{-3} μm), which is not so favourable for ink deposition to fabricate piezo-sensor circuits. The present study found that the solution-casting method is more economical and preferable for piezoelectric applications, particularly for PVDF-based composites.

References

[1] Sekhar MC, Veena E, Kumar NS, Naidu KCB, Mallikarjuna A, Basha DB. A review on piezoelectric materials and their applications. Crystal Research and Technology. 2023;58(2):2200130.
[2] Uchino K. The development of piezoelectric materials and the new perspective. In Advanced piezoelectric materials 2017 (pp. 1–92). Woodhead Publishing.
[3] Behera A. Piezoelectric materials. In Advanced materials: an introduction to modern materials science 2021 (pp. 43–76). Cham: Springer International Publishing.
[4] Jaffe B, Roth RS, Marzullo S. Piezoelectric properties of lead zirconate-lead titanate solid-solution ceramics. Journal of Applied Physics. 1954;25(6):809–10.
[5] Newnham RE. Electroceramics. Reports on Progress in Physics. 1989;52(2):123.
[6] Qi F, Xu L, He Y, Yan H, Liu H. PVDF-based flexible piezoelectric tactile sensors. Crystal Research and Technology. 2023;58(10):2300119.
[7] Khalifa M, Mahendran A, Anandhan S. Durable, efficient, and flexible piezoelectric nanogenerator from electrospun PANi/HNT/PVDF blend nanocomposite. Polymer Composites. 2019;40(4):1663–75.
[8] Fan FR, Tang W, Wang ZL. Flexible nanogenerators for energy harvesting and self-powered electronics. Advanced Materials. 2016;28(22):4283–305.
[9] Granstrom J, Feenstra J, Sodano HA, Farinholt K. Energy harvesting from a backpack instrumented with piezoelectric shoulderstraps. Smart materials and structures. 2007;16(5):1810.
[10] Joseph J, Kumar M, Tripathy S, Kumar GS, Singh SG, Vaniari SRK. A highly flexible tactile sensor with self-poled electrospun PVDF nanofiber. In 2018 IEEE SENSORS 2018, October (pp. 1–4). IEEE.
[11] Xin Y, Zhu J, Sun H, Xu Y, Liu T, Qian C. A brief review on piezoelectric PVDF nanofibers prepared by electrospinning. Ferroelectrics. 2018;526(1):140–51.

[12] Kitsara M, Blanquer A, Murillo G, Humblot V, Vieira SDB, Nogués C, ... Barrios L. Permanently hydrophilic, piezoelectric PVDF nanofibrous scaffolds promoting unaided electromechanical stimulation on osteoblasts. Nanoscale. 2019;11(18):8906–17.
[13] Chen L, Si Y, Zhu H, Jiang T, Guo Z. A study on the fabrication of porous PVDF membranes by in-situ elimination and their applications in separating oil/water mixtures and nano-emulsions. Journal of Membrane Science. 2016;520:760–68.
[14] El Achaby M, Arrakhiz FZ, Vaudreuil S, Essassi EM, Qaiss AJASS. Piezoelectric β-polymorph formation and properties enhancement in graphene oxide–PVDF nanocomposite films. Applied Surface Science. 2012;258(19):7668–77.
[15] Kumar SN, Kumar P. Dielectric and Pyroelectric Properties of Ferroelectric SBN–PVDF Composites Prepared by Hot-Uniaxial Press. International Journal of Applied Ceramic Technology. 2013;10: E11–E17.
[16] Rawat M, Yadav KL. Dielectric, enhanced magnetic and magnetodielectric properties of hot pressed (BNBT-BFO)/PVDF composite films. Journal of Polymer Research. 2015;22(12):230.
[17] Feng Y, Li JL, Li WL, Li ML, Chi QG, Zhang TD, Fei WD. Effect of BaTiO3 nanowire distribution on the dielectric and energy storage performance of double-layer PVDF-based composites. Composites Part A: Applied Science and Manufacturing. 2019;125:105524.
[18] Na H, Li Q, Sun H, Zhao C, Yuan X. Anisotropic mechanical properties of hot-pressed PVDF membranes with higher fiber alignments via electrospinning. Polymer Engineering & Science. 2009;49(7):1291–98.
[19] Yoo M, Frank CW, Mori S, Yamaguchi S. Interaction of poly (vinylidene fluoride) with graphite particles. 2. Effect of solvent evaporation kinetics and chemical properties of PVDF on the surface morphology of a composite film and its relation to electrochemical performance. Chemistry of materials. 2004;16(10):1945–53.
[20] Yang Y, Pan H, Xie G, Jiang Y, Chen C, Su Y, ... Tai H. Flexible piezoelectric pressure sensor based on polydopamine-modified BaTiO3/PVDF composite film for human motion monitoring. Sensors and Actuators A: Physical. 2020;301:111789.
[21] Naik R, Rao TS. Preparation and characterization of flexible PVDF based polymer film for energy harvesting applications. Materials Today: Proceedings. 2019;18:5107–13.
[22] Shanmugaraj P, Swaminathan A, Ravi RK, Dasaiah M, Senthil Kumar P, Sakunthala A. Preparation and characterization of porous PVdF-HFP/graphene oxide composite membranes by solution casting technique. Journal of Materials Science: Materials in Electronics. 2019;30(22):20079–87.
[23] Pramod K, Gangineni RB. Influence of solvent evaporation rate on crystallization of poly (vinylidene fluoride) thin films. Bulletin of Materials Science. 2015;38(4):1093–98.
[24] Gheibi A, Bagherzadeh R, Merati AA, Latifi M. Electrical power generation from piezoelectric electrospun nanofibers membranes: electrospinning parameters optimization and effect of membranes thickness on output electrical voltage. Journal of Polymer Research. 2014;21(11):571.
[25] Lee JE, Shin YE, Lee GH, Kim J, Ko H, Chae HG. Polyvinylidene fluoride (PVDF)/cellulose nanocrystal (CNC) nanocomposite fiber and triboelectric textile sensors. Composites Part B: Engineering.
[26] Nivedhitha DM, Jeyanthi S. Polyvinylidene fluoride, an advanced futuristic smart polymer material: A comprehensive review. Polymers for Advanced Technologies. 2023;34(2):474–505.
[27] Dallaev R, Pisarenko T, Sobola D, Orudzhev F, Ramazanov S, Trčka T. Brief review of PVDF properties and applications potential. Polymers. 2022;14(22):4793.
[28] Chu CC. Biotextiles as medical implants: 11. In Materials for absorbable and nonabsorbable surgical sutures 2013 Elsevier Inc. Chapters.
[29] Cheng Y, Li D. Numerical analysis of piezoelectric signal of PVDF membrane flapping wing in flight. In IOP Conference Series: Materials Science and Engineering 2020, March (Vol. 774, No. 1, p. 012090). IOP Publishing.

[30] Holmes-Siedle AG, Wilson PD, Verrall AP. PVdF: An electronically-active polymer for industry. Materials & Design. 1983;4(6):910–18.
[31] Liu R, Yuan B, Zhong S, Liu J, Dong L, Ji Y, ... He W. Poly (vinylidene fluoride) separators for next-generation lithium based batteries. Nano Select. 2021;2(12):2308–45.
[32] Barwar A, Kala P, Singh R. 3D-printed conformal sensors for health monitoring of bovine post diaphragmatic hernia surgery: a review. Rapid Prototyping Journal. 2024a;30(5):904–18.
[33] Huang J, Di Silvio L, Wang M, Tanner KE, Bonfield W. In vitro mechanical and biological assessment of hydroxyapatite-reinforced polyethylene composite. Journal of Materials Science: Materials in Medicine. 1997;8(12):775.
[34] Mi FL, Shyu SS, Wu YB, Lee ST, Shyong JY, Huang RN. Fabrication and characterization of a sponge-like asymmetric chitosan membrane as a wound dressing. Biomaterials. 2001;22(2):165–73.
[35] Ranjan N, Singh R, Ahuja IPS. Development of PLA-HAp-CS-based biocompatible functional prototype: a case study. Journal of Thermoplastic Composite Materials. 2020b;33(3):305–23.
[36] Singh S, Singh G, Prakash C, Ramakrishna S, Lamberti L, Pruncu CI. 3D printed biodegradable composites: An insight into mechanical properties of PLA/chitosan scaffold. Polymer Testing. 2020;89:106722.
[37] Suyatma NE, Copinet A, Tighzert L, Coma V. Mechanical and barrier properties of biodegradable films made from chitosan and poly (lactic acid) blends. Journal of Polymers and the Environment. 2004;12(1):1–6.
[38] Barwar A, Kala P, Singh R, Davim JP. 3D printed hernia mesh implant: a conformability study. Exploration of BioMat-X. 2024b;1(5):266–79.
[39] Zhou Y, Chen CC, Volakis JL. Dual band proximity-fed stacked patch antenna for tri-band GPS applications. IEEE Transactions on Antennas and Propagation. 2007;55(1):220–23.
[40] Barwar A, Kala P, Singh R. On the design and fabrication of polyvinylidene fluoride-based sensors to detect recurrence of diaphragmatic hernia. Proceedings of the Institution of Mechanical Engineers, Part C: Journal of Mechanical Engineering Science. 2025;239(8):2681–99.
[41] Ranjan N, Singh R, Ahuja IPS, Rahman M, Ramakrishna S. PLA-HAp-CS-based biocompatible scaffolds prepared through micro-additive manufacturing: a review and future applications. 3D printing in Biomedical Engineering. 2020a;209–29.
[42] Cai X, Lei T, Sun D, Lin L. A critical analysis of the α, β and γ phases in poly(vinylidene fluoride) using FTIR. RSC advances. 2017;7(25):15382–89.
[43] Nandiyanto ABD, Oktiani R, Ragadhita R. How to read and interpret FTIR spectroscope of organic material. Indonesian journal of science and technology. 2019;4(1):97–118.

Minhaz Husain*, Rupinder Singh, Gurwinder Singh

Chapter 2
Formulation of piezoelectric materials for 3D printing applications

Abstract: Over the past two decades, considerable work has been reported on piezoelectric materials; however, little has been reported on piezoelectric materials for three-dimensional (3D) printing applications. This chapter reports on the formulation and fundamental working of piezoelectric materials for 3D printing applications. Piezoelectric materials are now crucial for smart, energy-responsive, and self-sensing applications, with 3D printing being a key method for incorporating them into functional devices. In this study, the development and performance of piezoelectric composites are considered in the context of additive manufacturing. Different polymeric and ceramic matrices were considered, including polyvinylidene fluoride (PVDF), barium titanate ($BaTiO_3$), and lead zirconate titanate (PZT), among others, together with nano-reinforcements such as carbon nanotubes (CNTs), glass fibers (which also contain an additional ferroelectric phase, PVDF, and trifluoroethylene), or organic fillers to improve electromechanical response and structural integrity. Their mechanical properties, in terms of compressive, tensile, and flexural strength, were determined and discussed in reference to the available literature reported for both virgin and reinforced materials. The research indicates that piezoelectric ceramics can be evenly distributed within a polymer matrix, thereby enhancing β-phase crystallinity and increasing the piezoelectric coefficient (d_{33}).

Keywords: 3D printing, piezoelectric materials, formulation, piezoelectric coefficient

2.1 Introduction

Developments in the formulation of piezoelectric materials for 3D printing applications are rapidly advancing the frontier between materials science, additive manufacturing, and smart systems engineering [1]. Piezoelectric materials, such as lead zirconate titanate (PZT) ceramics and polyvinylidene fluoride (PVDF) polymers, were traditionally produced using conventional processing methods, including sintering, film casting, or extrusion [1–3]. Still, the emergence of 3D printing in their manu-

***Corresponding author: Minhaz Husain**, Mechanical Engineering Department, Marwadi University Research Center, Marwadi University, Rajkot, Gujarat, India, e-mail: minhazme2012@gmail.com
Rupinder Singh, Gurwinder Singh, Mechanical Engineering Department, National Institute of Technical Teachers Training and Research, Chandigarh, India

 | https://doi.org/10.1515/9783112218846-002

facture has transformed how these materials can be manufactured, integrated, and utilized [4]. Additive manufacturing also enables the fabrication of complex, embedded, and graded composite architectures that are not possible with conventional techniques. The depicted transition has enabled the direct 3D printing of piezoelectric structures into devices themselves, thereby eliminating assembly steps, offering increased design freedom, and facilitating integrated sensing, actuation, and energy harvesting with custom shapes [5].

2.1.1 Fundamentals of piezoelectric materials

Figure 2.1(a) shows that the molecules of piezoelectric materials are randomly distributed because no pressure or force is applied. Figure 2.1(b) depicts that the force is applied to the piezoelectric materials, the molecules of piezoelectric materials are arranged in order, and charge or voltage is generated during the shifting of positive and negative ions. The charge generated during applied mechanical load (*F*), the cross-sectional area (*A*) of the piezoelectric materials, and the thickness of the piezoelectric materials (*t*) are given as follows [6]:

$$q = D_{33} \times F \tag{2.1}$$

where D_{33} and E_{33} are the piezoelectric constants, and the generated voltage (*V*) is given by (2.2) [7]

$$V = \frac{D_{33}t}{E_{33}A} \times F \tag{2.2}$$

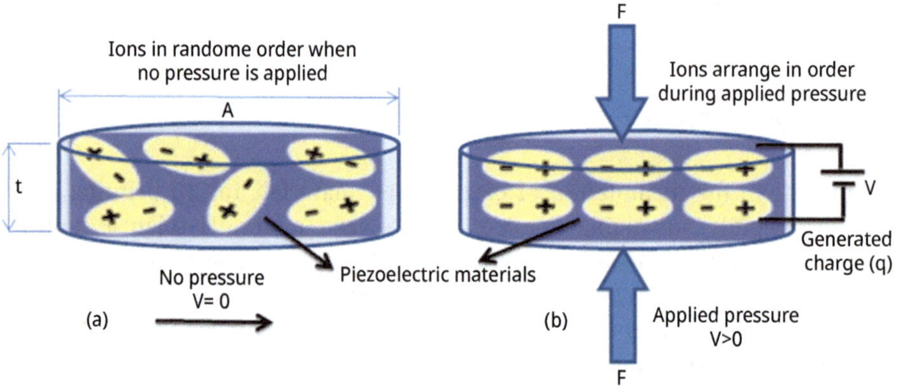

Figure 2.1: Working of piezoelectric materials: (a) when pressure is not applied and (b) when pressure is applied.

2.1.2 3D printing of piezoelectric materials

In 3D-printed piezoelectric devices, the materials of choice are often thermoplastic piezoelectric polymers, such as PVDF or its copolymers (e.g., polyvinylidene fluoride-trifluoroethylene (PVDF-TrFE)), which are frequently filled and/or reinforced with ceramic particles, including barium titanate ($BaTiO_3$) or PZT [8]. For example, SLS has been utilized to print PVDF/$BaTiO_3$ composites with high ceramic loading and uniform dispersion, thereby achieving strong β-phase crystallinity and effective piezoelectricity [9]. These formulation approaches utilize polymer matrices that can be readily processed and formed, in combination with the high electromechanical coupling of ceramic fillers, to produce 3D-printed sensors, actuators, and energy scavengers. In another example, the fused filament fabrication (FFF) of PVDF has been optimized by stretching and poling after printing to increase the β-phase fraction [10]. One study demonstrated a d_{33} (piezoelectric charge coefficient) of 7.29 pC/N in printed films, which is significantly better than that of previously printed PVDF. A 3D-printed piezoelectric scaffold was also demonstrated for biomedical applications, such as bone defect repair, in which the coupled electrospun PVDF nanofibers, aligned within a 3D-printed structure, operated via self-powered stimulation via mechanical deformation [11, 12].

The revolution brought about by 3D printing in piezoelectric material production lies not only in the freedom of fabricating functional parts, but also in reinforcement strategies that leverage both multi-material printing and bone regeneration to fine-tune and adapt mechanical and piezoelectric properties [13]. In both industry and academia, piezoelectric polymers are reinforced with nanoparticles, nanogenerators, nanofibers, or micro-sized fillers to enhance their functional performance [14]. Notably, carbon nanotubes (CNTs) have become increasingly attractive as a reinforcing phase due to their exceptionally high tensile strength and modulus [15]. The presence of multiwalled CNTs (MWCNTs) in PVDF/$BaTiO_3$/CNT composites provides nucleation sites for the β-phase of PVDF [16], thereby enhancing electrical connectivity, improving piezoelectric response, and increasing mechanical strength [17]. Other conductive carbon-based fillers, such as graphene nanoplatelets and reduced graphene oxide, have also been incorporated into electrospun PVDF fibers to form reinforced membranes with enhanced mechanical strength and piezoelectric response [18].

Another type of reinforcement is glass fiber or organic fiber. Even fewer studies specifically report the flexural or tensile strength of their 3DP piezocomposites reinforced with glass. Therefore, we can only speculate on these values, drawing from general composite literature. For instance, commonly encountered fiber-glass-reinforced polymers, such as those featured in the range of hundreds of MPa and Young's moduli significantly higher than those of plain polymers [19]. In piezoelectric composites, a combination of glass fibers and PVDF or thermosetting resins with ceramic piezoparticles may theoretically achieve stiffness and structural strength while maintaining piezoelectric properties. Thermosetting resins, such as liquid crystalline resins, have been employed as matrix materials for PZT composites, which can then be processed at high temperatures while retaining good mechanical toughness and piezoelectric coefficient values [20].

Looking at the intrinsic mechanical performance of the "virgin" piezoelectric materials (unreinforced ceramics or polymers), these, then, serve as reference values against which composite systems have to be compared. For instance, it is known that even commercial PZT ceramics (the type generally used in actuators) have been investigated under mechanical loads: four-point bending tests demonstrate that monolithic PZTs have a flexural (bending) strength which can be best understood with the aid of Weibull statistics, as well its fracture toughness has been estimated using single-edge notch beam techniques [21]. Under compressive stress, piezoelectric ceramics can withstand peak loads; for example, it is reported that PZT can withstand a dynamic compression load of ~69–103 MPa in sonar transducer operation, which places severe limitations on device mechanical design. In recent years, attempts to further increase the strength (thereby durability) of ceramics have led to compressive stress capacity even higher; for example, PZT ceramics of advanced strength have shown an ultimate compressive stress limit approaching ~1,450 MPa under well-controlled conditions [22].

On the polymer side, PVDF exhibits quite different mechanical behavior. Whereas ceramics are at the stiff end of the spectrum, brittle and subject to fatigue, PVDF is flexible and ductile [23]. It is reported that the tensile strength, elongation at break, and fatigue resistance of PVDF film are good due to its semicrystalline nature and processing-induced β-phase crystallinity. Although the absolute value depends on the specific processing (extrusion, stretching, poling, etc.), typical printed PVDF structures already possess sufficient mechanical strength for sensing and actuation. For instance, 3D-printed PVDF can retain satisfactory mechanical strength in selective laser sintering or FFF, serving as energy-harvesting films or flexible sensors [24].

When such virgin piezomaterials are filled (e.g., PVDF/$BaTiO_3$ composites printed by FFF), their mechanical properties (e.g., compression strength) and electromechanical coupling increase significantly compared to their respective monophase systems [25, 26]. An investigation into the PVDF/$BaTiO_3$ 3D porous foam showed that, in addition to higher piezoelectric output, it had a higher compressive strength than pure PVDF foams. The ceramic phase, which provides internal reinforcement and stress transfer, results in a high elongation at yield under load for the porous ligament architecture [27]. Moreover, for PZT/polymer composite, the PZT volume fraction and forming pressure used in the preparation of the composites also have a significant influence on the mechanical modulus and d_{33}; increasing the volume fraction of PZT increases stiffness as well as piezoresponse, but meanwhile causes more brittleness; accordingly, one needs to be careful during processing.

In the case of CNT reinforcement, for example, research on PVDF/$BaTiO_3$/MWCNT films indicates that MWCNTs not only enhance piezoelectric performance but also augment mechanical properties as follows: the stress applied to the polymer network is distributed by nanotubes, and resilience to cyclic deformation is improved [28]. Another approach to forming PVDF-HFP (polyvinylidene fluoride-co-hexafluoropropylene) (as a copolymer) nanocomposites with $BaTiO_3$ and CNTs has demonstrated high tensile strength and flexibility for 3D printing, as well as poling at High Fields to provide a potent piezoelectric response [29]. These results reveal that reinforcement with nanostructured fillers, such as CNTs, is efficient not only for enhancing mechanical properties but also for improving the performance of printed piezoelectric composites.

In terms of industrial formulation paradigms, these considerations can be translated into design rules: for 3D printing piezoelectric devices, select a printable base polymer (such as PVDF), add ceramic or nanofiller reinforcement to expose higher piezoelectric output, and balance the filler content in such a way that it is printable while also maintaining acceptable mechanical reliability and electromechanical performance. The processing conditions (temperature, stretching, and poling) must be adjusted to increase the β-phase (the active dipolar phase) without affecting the matrix. Several extrusion/granulation cycles (for filament), homogeneous filler distribution, and high-field poling (typically thermal poling) are required in industrial workflows to obtain composite filaments or resin-based materials with reproducible characteristics [30].

Piezoelectric 3D-printed composites are utilized in research to construct innovative scaffolds, wearable sensors, self-powered energy harvesters, and integrated actuators. Porosity in PVDF/$BaTiO_3$ scaffolds has been introduced using selective laser sintering or through foaming steps to demonstrate effective energy harvesting and mechanical deformation sustainability, illustrating how the co-design of formulation and architecture results in multifunctionality [11, 12, 31]. In soft robotics, the inkjet or extrusion printing of PVDF-TrFE/CNT composites enables the fabrication of flexible actuators and sensors. In biomedicine, patterns of aligned PVDF nanofibers within 3D-printed substrates provide mechano-electrical cues to cells, facilitating tissue repair.

The design of 3D piezoelectric materials for 3D printing leverages the unique features of additive manufacturing (namely, design freedom, integration of multiple materials, and geometric complexity) to develop new piezoelectric materials with specific functional and mechanical characteristics [32]. By using reinforcement strategies (specifically ceramic particles, CNTs, and polymer blends), researchers can increase piezoelectric output without compromising durability [33]. The as-received mechanical behavior of the virgin piezomaterials (e.g., compressive strength for ceramic and tensile strength for PVDF) dictates how much reinforcement is required, as well as the processing of the materials. Guided by reported literature data, such as the compressive stress limit of PZT (~69–103 MPa in dynamic conditions), and the increased compressive strength achieved with PVDF/$BaTiO_3$ foams, helps inform formulation decisions. Remove the dithering, and you get a pony: a rich and expanding toolkit for designing functional, rugged, application-specific 3D-printed piezoelectric systems [34].

Piezo is short for the Greek word for pressure, and that's the idea: squeeze certain materials, and they generate an electric charge. It is known as the pressure-to-electricity behavior, also referred to as piezoelectricity, which works in reverse:

changing shape in response to an applied voltage [35]. Natural and synthetic piezoelectric materials encompass both naturally occurring, mineral-based materials, such as quartz, and artificial materials, including synthesized ceramic PZTs or PVDFs, which share a typical crystal structure that lacks internal electric dipoles. When you hear about piezoelectricity today, you often hear about it in the same breath as 3D printing, because additive manufacturing enables these materials to be shaped into forms that could not previously have been cut, pressed, or molded. Instead of planar films or dense ceramic blocks, you can print lattices, curved sensors, flexible strips, and composite structures with controlled filler placement [36]. This opens the possibility that a piezoelectric layer isn't just affixed to a device but is actually developed within it. In 3D printing, piezoelectric materials are currently used for embedded sensors, vibration energy harvesters, soft robotics actuators, microfluidic pumps, and wearable devices. The true benefit is that designers can adjust geometry, stiffness, and material composition simultaneously to create custom-designed, functional structures that respond to their surroundings [37].

2.2 Materials and methods

The development of piezoelectric materials begins with selecting an appropriate matrix and functional phase, as printability versus actual performance is determined by how well these components mix, disperse, and process in the base matrix. In the majority of lab workflows, the underlying matrix is a polymer, such as PVDF or epoxy, or a ceramic system based on PZT or $BaTiO_3$. Working of piezoelectric materials under compressive load is shown in Figure 2.2. The powders or pellets are typically measured and prepared first, then dried to a moisture-free state, and, if possible, mixed with additives that also modify viscosity, particle size distribution, or flow. Such ceramic-filled systems typically require ball milling or high-shear mixing to break down agglomerates and ensure an even coating within the matrix, as clumps of this material can clog nozzles and reduce electrical output.

2.2.1 Polymer-based formulations can be more easily tuned

The polymer is dissolved or melted, and the piezoelectric powder is slowly mixed in. Dispersing agents (or surfactants) prevent the mixture from settling. Once the mix is uniform, it's tailored to the specific 3D printing process. For direct ink writing, the pastes require high viscosity and thixotropic characteristics, whereas for FFF, pellets must be extruded without collapsing or burning. In resin-based printing, UV-curable composites are required to keep the piezoelectric particles suspended and prevent them from interfering with the polymerization process. And after you print, almost

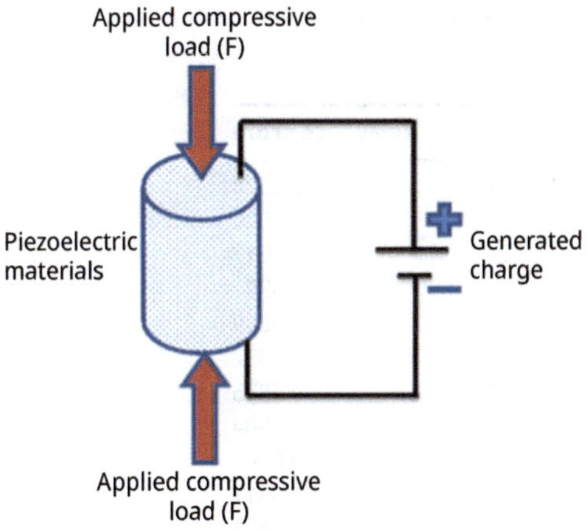

Figure 2.2: Working of piezoelectric materials.

everything still requires post-processing. Ceramics are sintered at high temperatures to freeze in density and grain structure. Polymers can be stretched or annealed to increase the β-phase content and enhance their piezoelectric properties. Nearly all systems must be electrically poled, i.e., subjected to a high applied electric field, so that the material's dipoles align. The target at each step, he says, remains consistent: design a formulation that prints cleanly and holds together well while producing the desired piezoelectric response.

2.3 Bibliometric analysis for the piezoelectric material fabricated by 3D printing

The bibliometric analysis was performed using the Scopus database over the last 20 years (as of January 2026), with the keyword "piezoelectric materials fabricated by 3D printing," yielding 139 documents, including research articles and conference papers. These results were further sorted by citation from highest to lowest, and a .csv file was exported for the development of a bibliometric network diagram using the open-source VOSviewer software. The .csv file was then opened using the VOS-viewer software. At the 5-level occurrences, a total of 4,700 terms were observed. Out of these, 168 terms met the threshold, and from this, the top 60% terms were selected for the bibliomeric network diagram, as shown in Figure 2.3. Then, 100 terms were selected based on occurrence and relevance scores (Table 2.1).

Table 2.1: Occurrences and relevance score for the keywords "piezoelectric materials fabricated by 3D printing."

Id	Term	Occurrences	Relevance score
1	3D printing technology	11	0.6574
2	3D structure	7	0.6029
3	Actuator	12	0.8317
4	Addition	13	0.432
5	Advantage	15	0.5669
6	Array	7	0.5532
7	Barium titanate	14	0.6835
8	Batioinf3	9	0.6604
9	Batioinf3 inf	13	0.856
10	Binder	6	1.9817
11	Bone regeneration	6	2.0321
12	Bto	9	0.9805
13	Carbon nanotube	7	0.7416
14	Cell	11	0.5587
15	Ceramic	20	0.8255
16	Change	11	0.5392
17	Combination	11	0.4238
18	Construction	5	0.7213
19	Cost	9	0.7723
20	Defect	11	0.9266
21	Demand	9	0.6308
22	Deposition	11	0.6882
23	Design methodology approach	5	2.2667
24	Diameter	10	0.7334
25	Dielectric property	6	2.2761
26	Digital light processing	5	3.3721
27	Dimensional printing	5	0.7346
28	Dinf33 inf	6	1.3341
29	DIW	5	0.709
30	DLP	5	3.3721
31	Electrical property	6	0.9071
32	Energy harvester	14	0.6112
33	Experimental result	9	0.6014
34	Extrusion	12	0.7157
35	Fabrication process	8	0.4397
36	Factor	15	1.2069
37	FDM	8	0.4923
38	Feasibility	8	0.5054
39	FFF	6	1.6209
40	Filament fabrication	6	1.6209
41	Filler	11	0.5008
42	First time	5	0.6495
43	Flexible electronic	5	0.7487
44	Force	13	0.7042
45	Form	8	0.7243

Table 2.1 (continued)

Id	Term	Occurrences	Relevance score
46	Frequency	17	0.5168
47	Function	9	0.4983
48	High sensitivity	6	0.4917
49	Impact	5	1.0473
50	Influence	15	0.4539
51	Lead zirconate titanate	10	1.3077
52	Load	10	0.4386
53	Low cost	11	1.1169
54	Mems	8	1.2552
55	Microstructure	7	0.6614
56	MPa	5	1.1401
57	Multi material	6	1.0027
58	Need	12	0.8563
59	Numerical simulation	6	1.8455
60	Order	7	0.4572
61	Originality value	5	2.2667
62	Output voltage	9	0.5523
63	PCN	22	0.6636
64	PDMS	8	0.9082
65	Peak	8	0.5502
66	Phase	10	0.3459
67	Piezoelectric	6	1.4647
68	Piezoelectric ceramic	9	1.0583
69	Piezoelectric coefficient	16	0.9177
70	Piezoelectric sensor	14	0.4745
71	Piezoelectric transducer	6	1.4204
72	Piezoelectricity	11	0.505
73	PLA	8	0.773
74	Poly	19	0.6447
75	Polydimethylsiloxane	8	0.9082
76	Polylactic acid	5	1.496
77	Polyvinylidene fluoride	9	0.6631
78	Possibility	8	1.1708
79	Preparation	5	2.5673
80	Pressure	10	0.5614
81	Printer	12	0.9783
82	Prototype	7	0.6323
83	PVDF	15	0.6234
84	Range	23	0.5563
85	Resolution	9	1.4489
86	Resonant frequency	8	1.5947
87	Scaffold	13	0.8126
88	Self	10	0.6062
89	SEM	8	0.4437
90	Sensitivity	16	0.4395
91	SLA	5	3.9157

Table 2.1 (continued)

Id	Term	Occurrences	Relevance score
92	Stereolithography	11	2.3757
93	Temperature	12	0.6968
94	Term	7	0.6912
95	Variety	6	1.785
96	Vinylidene fluoride	9	1.0813
97	Vol	5	1.3051
98	Way	8	0.6315
99	Wearable device	6	1.0648
100	Weight	5	0.8019

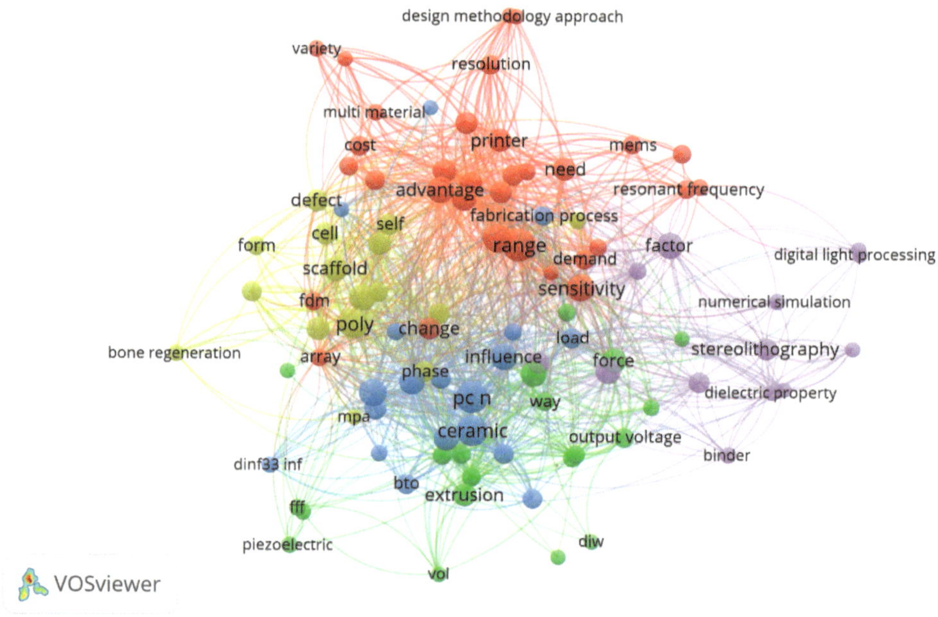

Figure 2.3: Bibliometric network diagram for the keywords "piezoelectric materials fabricated by 3D printing."

The research gap analysis was conducted for the keyword "piezoelectric coefficient," and the reported work is highlighted in Figure 2.4. From this work, no keywords related to 3D printing and the piezoelectric coefficient were identified.

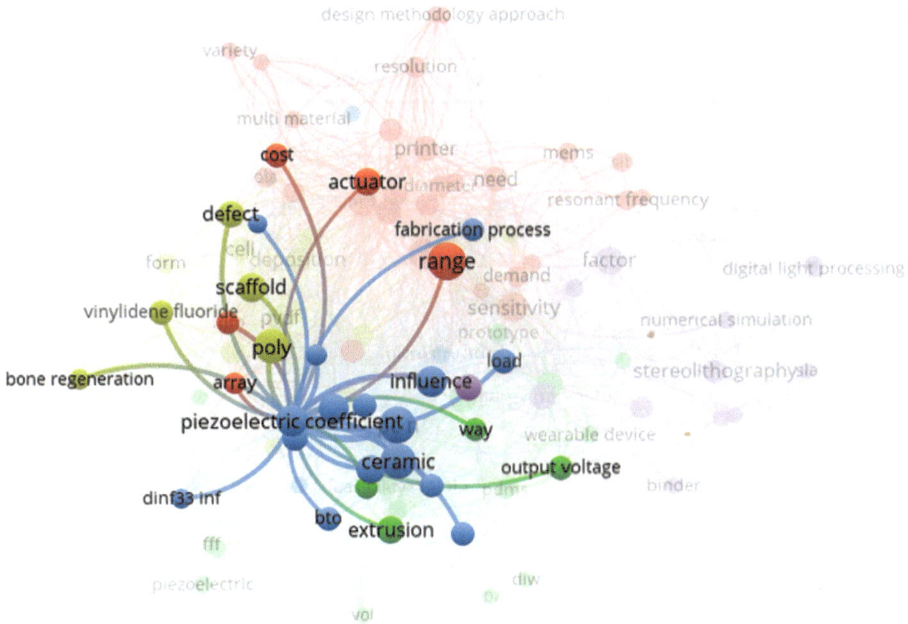

Figure 2.4: Research gap for the keywords piezoelectric coefficient.

The other keywords were also explored in the Scopus database using the search term "formulation of piezoelectric materials for the 3D printing application," and only 22 documents were reported (as of January 8, 2026). Then, the .csv file was exported to develop the bibliometric network diagram using the VOSviewer software, as shown in Figure 2.5, based on the occurrences and relevance scores for the keywords "formulation of piezoelectric materials for the 3D printing application," as listed in Table 2.2.

Table 2.2: Occurrences and relevance score for the keywords "formulation of piezoelectric materials for the 3D printing application."

Id	Term	Occurrences	Relevance score
1	Additive manufacturing	5	0.6895
2	Binder	3	2.1551
3	Challenge	7	0.6448
4	Cycle	4	0.6937
5	Dielectric property	4	1.6059
6	Direct ink writing	3	1.6789
7	Dispersant	4	1.6264
8	DIW	3	1.6789
9	Effect	5	0.6127
10	Energy harvester	5	0.5487

Table 2.2 (continued)

Id	Term	Occurrences	Relevance score
11	Factor	3	1.3642
12	Functionality	3	1.4928
13	Ink	7	0.3724
14	Ink formulation	4	1.2952
15	Lead zirconate titanate	4	0.8298
16	Optimization	4	1.4052
17	PCN	5	1.2506
18	Performance	6	0.6475
19	Piezoelectric material	3	0.7679
20	Piezoelectric property	6	0.6025
21	Polyvinylidene fluoride	4	0.6998
22	Printing	17	0.0758
23	SLA	3	0.1723
24	Technology	7	0.5682
25	Use	4	1.9062
26	Vol	3	0.6151

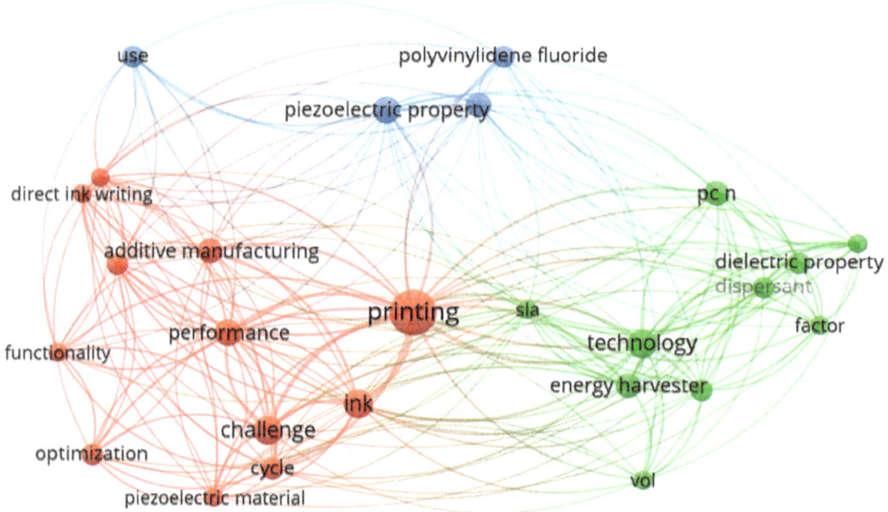

Figure 2.5: Bibliometric network diagram for the keywords "formulation of piezoelectric materials for the 3D printing application."

The research gap analysis was performed for the keyword "piezoelectric property," and the reported work is highlighted in Figure 2.6. From this work, no keywords related to 3D-printed piezoelectric properties were identified.

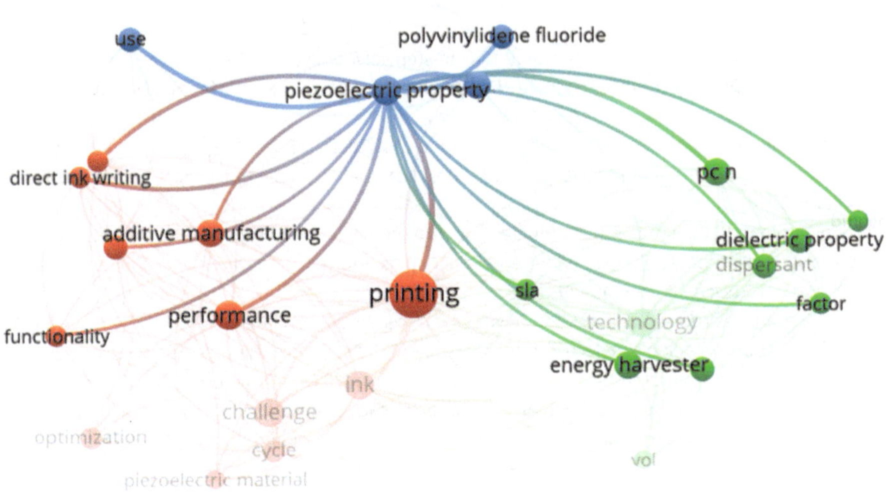

Figure 2.6: Research gap for the keywords piezoelectric property.

2.4 Summary

The study verifies that 3D printing offers a new route to completely transform the process for piezoelectric materials, enabling the fabrication of complex, controllable, and functionally graded structures. The design of efficient piezoelectric composites is challenging, requiring careful trade-offs among printability, mechanical strength, and piezoelectric performance through control of filler type, matrix dispersion, and alignment. Virgin PVDF and PZT have baseline properties suitable for use as structural materials, with only glass fibers, CNTs, or organic reinforcements. These reinforcements enhance compressive and tensile strength without limiting piezoelectric output. The higher β-phase of PZT helped impart endurance response and charge generation in the printed composites under cyclic stress. This work demonstrates that synergistic additive combination and material design can be used to create lightweight, flexible, and functional piezoelectric devices for industrial sensors and wearable electronics.

References

[1] Bodkhe S, Turcot G, Gosselin FP, Therriault D. One-step solvent evaporation-assisted 3D printing of piezoelectric PVDF nanocomposite structures. ACS applied materials & interfaces. 2017 Jun 21; 9 (24):20833–42.

[2] Husain M, Singh R, Pabla BS. On secondary recycling of PVDF for 3D printing of functional sensors: mechanical, flow, thermal, and dielectric properties. National Academy of Sciences Letters. 2024a; 47(4):385–90.

[3] Husain M, Singh R, Pabla BS. Effect of meta-structure on mechanical properties of polyvinylidene fluoride composite-based 3D printed intramedullary pins. Journal of Thermoplastic Composite Materials. 2024b Oct; 37(10):3319–39.

[4] Kim H, Renteria-Marquez A, Islam MD, Chavez LA, Garcia Rosales CA, Ahsan MA, Tseng TL, Love ND, Lin Y. Fabrication of bulk piezoelectric and dielectric BaTiO3 ceramics using paste extrusion 3D printing technique. Journal of the American Ceramic Society. 2019 Jun; 102(6):3685–94.

[5] Ingaglio J, Fox J, Naito CJ, Bocchini P. Material characteristics of binder jet 3D printed hydrated CSA cement with the addition of fine aggregates. Construction and Building Materials. 2019 May 10; 206:494–503.

[6] De Marzo G, Fachechi L, Antonaci V, Mastronardi VM, Portaluri L, Todaro MT, Algieri L, Qualtieri A, Rizzi F, Scaraggi M, De Vittorio M. On the measurement of piezoelectric d33 coefficient of soft thin films under weak mechanical loads: A rapid and affordable method. Materials & Design. 2024 November 1; 247:113399.

[7] Ramesh S, Ravinder D, Naidu KB, Kumar NS, Srinivas K, Basha DB, Sekhar BC. A review on giant piezoelectric coefficient, materials, and applications. Biointerface Research in Applied Chemistry. 2019 Oct; 9(5):4205–16.

[8] Chang SM, Hur S, Park J, Lee DG, Shin J, Kim HS, Song SE, Baik JM, Kim M, Song HC, Kang CY. Optimization of piezoelectric polymer composites and 3D printing parameters for flexible tactile sensors. Additive Manufacturing. 2023 April 5; 67:103470.

[9] Vadodaria S, Mills T. Jetting-based 3D printing of edible materials. Food Hydrocolloids. 2020 September 1; 106:105857.

[10] Husain M, Singh R, Pabla BS. On process capability of single screw extruder for fabricating PVDF composite matrix. Materials Today: Proceedings. 2023 November 10.

[11] Tao R, Granier F, Therriault D. Multi-material freeform 3D printing of flexible piezoelectric composite sensors using a supporting fluid. Additive Manufacturing. 2022 December 1; 60:103243.

[12] Tao R, Shi J, Rafiee M, Akbarzadeh A, Therriault D. Fused filament fabrication of PVDF films for piezoelectric sensing and energy harvesting applications. Materials Advances. 2022; 3(12):4851–60.

[13] Donate R, Paz R, Moriche R, Sayagués MJ, Alemán-Domínguez ME, Monzón M. An overview of polymeric composite scaffolds with piezoelectric properties for improved bone regeneration. Materials & Design. 2023 Jul 1; 231:112085.

[14] Parida K, Thangavel G, Cai G, Zhou X, Park S, Xiong J, Lee PS. Extremely stretchable and self-healing conductor based on thermoplastic elastomer for all three-dimensional printed triboelectric nanogenerator. Nature Communications. 2019 May 14; 10(1):2158.

[15] Simunec DP, Breedon M, Muhammad FU, Kyratzis L, Sola A. Electrical capability of 3D printed unpoled polyvinylidene fluoride (PVDF)/thermoplastic polyurethane (TPU) sensors combined with carbon black and barium titanate. Additive Manufacturing. 2023 Jul 5; 73:103679.

[16] Husain M, Singh R, Pabla BS. On Prominent β-Phase of PVDF Matrix with 4D Capabilities. Journal of The Institution of Engineers (India): Series C. 2025; 106(1):341–52.

[17] Hossain SS, Jang S, Park S, Bae CJ. Understanding ink design and printing dynamics of extrusion-based 3D printing: Defect-free dense piezoelectric ceramics. Journal of Manufacturing Processes. 2023 April 28; 92:1–1.

[18] García-Farrera B, Velásquez-García LF. Ultrathin ceramic piezoelectric films via room-temperature electrospray deposition of ZnO nanoparticles for printed GHz devices. ACS applied materials & interfaces. 2019 Aug 5; 11(32):29167–76.
[19] Damamme R, Seveyrat L, Borta-Boyon A, Nguyen VC, Le MQ, Cottinet PJ. 3D printing of doped barium-titanate using robocasting-Towards new generation lead-free piezoceramic transducers. Journal of the European Ceramic Society. 2023 Jul 1; 43(8):3297–306.
[20] Malik U, Selvakannan PR, Mazur M, Li Y, Bhargava SK. Robocasting – printing ceramics into functional materials. In Additive Manufacturing for Chemical Sciences and Engineering 2022 September 2 (pp. 109–36). Singapore: Springer Nature Singapore.
[21] Kainz M, Haudum S, Guillén E, Brüggemann O, Höller R, Frommwald H, Dehne T, Sittinger M, Tupe D, Major Z, Stubauer G. 3D Inkjet Printing of Biomaterials with Solvent-Free, Thiol-Yne-Based Photocurable Inks. Macromolecular Materials and Engineering. 2024 May; 309(5):2400028.
[22] Massaro A, Nguyen KG, Suárez DV, Glukharev A, Ingrosso C, Striccoli M, Kirlioglu AC, Kellici S, Middelkoop V. 3D printed C 3 N 4-based structures for photo-, electro-chemical and piezoelectric applications. Materials Chemistry Frontiers. 2025; 9(18):2730–43.
[23] Tilve-Martínez D, Neri W, Lessire J, Dulucq B, Vukadinovic N, Berton B, Yuan J, Poulin P. Ultralow UV absorber content in 3D printed nanocomposites: maximizing printability and microwave absorption efficiency. Journal of Physics: Materials. 2025 Jan 22; 8(1):015010.
[24] Song X, Chen Z, Lei L, Shung K, Zhou Q, Chen Y. Piezoelectric component fabrication using projection-based stereolithography of barium titanate ceramic suspensions. Rapid Prototyping Journal. 2017 Jan 16; 23(1):44–53.
[25] Gong G, You Y, Shen H, Laghaei M, Wang Y, Li Y. 3D printing piezoelectric materials: Innovations, challenges, and future perspectives. Materials Today Sustainability. 2025 Nov 19; 101257.
[26] Kumar R, Kumar V, Husain M, Singh R, Kumar P. On Fabrication of Novel Polyvinylidene Fluoride BaTiO3-Cu-Doped ZnO-Based Filament for 3D Printing of Pacemakers. Journal of Materials Engineering and Performance. 2025 Jul 7; 1–4.
[27] Carravero Costa L, Pochard I, Buron CC, Jurin FE. Elaboration of Conductive Hydrogels by 3D Printer for the Development of Strain Sensors. Gels. 2025 Jun 20; 11(7):474.
[28] Upadhye S, Bagde P, Sange SR, Rokade A, Chandak MA, Patil TR, Kakade NT, Shelke N. Optimization and multi-functional predictive performance of advanced superhydrophobic cementitious materials for water-resistant infrastructure. Asian Journal of Civil Engineering. 2025 May; 26(5):2023–36.
[29] Pazourek J, Pražák O, Procházka R, Straka L, Veselý P. Fabrication of Piezoelectric Composite Polymers Using Stereolithography. In 2025 International Spring Seminar on Electronics Technology (ISSE) 2025 May 14 (pp. 1–6). IEEE.
[30] Song S, Li Y, Wang Q, Zhang C. Facile preparation of high loading filled PVDF/BaTiO 3 piezoelectric composites for selective laser sintering 3D printing. RSC advances. 2021; 11(60):37923–31.
[31] Li B, Ma Y, Fatima K, Zhou X, Gu X, Chen S, He C. 3D printed shape-memory piezoelectric scaffolds with in-situ self-power properties for bone defect repair. Journal of Nanobiotechnology. 2025 Mar 24; 23(1):244.
[32] Lin X, Yu F, Zhang X, Li W, Zhao Y, Fei X, Li Q, Yang C, Huang S. Wearable piezoelectric films based on MWCNT-BaTiO3/PVDF composites for energy harvesting, sensing, and localization. ACS Applied Nano Materials. 2023 Jun 21; 6(13):11955–65.
[33] Van den Ende DA, De Almeida P, Van der zwaag S. Piezoelectric and mechanical properties of novel composites of PZT and a liquid crystalline thermosetting resin. Journal of Materials Science. 2007 Aug; 42(15):6417–25.
[34] Bermejo R, Deluca M. Mechanical characterization of PZT ceramics for multilayer piezoelectric actuators. Journal of Ceramic Science and Technology. 2012 Dec 1; 3(4):159–68.

[35] Butler JL, Rolt KD, Tito FA. Piezoelectric ceramic mechanical and electrical stress study. Journal of the Acoustical Society of America. 1994 September 1; 96(3):1914–17.
[36] Xi J, Chen H, Peng X, Wan H, Lin Y, Tan Z, Xing J, Chen Q, Zhu J, Ye ZG. Achieving significantly enhanced piezoelectricity in aurivillius ceramics by improving initial polarization and dielectric breakdown strength. Journal of the European Ceramic Society. 2023 Sep 1; 43(11):4757–65.
[37] Ahbab N, Naz S, Xu TB, Zhang S. A Comprehensive Review of Piezoelectric PVDF Polymer Fabrications and Characteristics. Micromachines. 2025 Mar 28; 16(4):386.

Gurwinder Singh*, Rupinder Singh, Sourav Kumar

Chapter 3
Material extrusion-based 3D printing for the manufacturing of piezoelectric functional prototypes

Abstract: In the past three decades, numerous studies have been reported on the fabrication of functional prototypes via the material extrusion (MEX)-based 3D printing for various engineering applications. However, little has been reported on the 3D printing of functional piezoelectric prototypes using MEX for biomedical applications. This chapter highlights the fabrication of tensile specimens (per ASTM D638 Type V) of thermoplastic polyurethane (TPU) on a 75% cotton-25% lycra fabric using a design of experiments approach. The process parameters for the fused filament fabrication/MEX process were determined to monitor muscle movement under tensile loading conditions. The results suggest that the sample, fabricated at 0° orientation, 100% infill density, and a 20-mm prestrain condition on a 75% cotton-25% lycra fabric, showed the highest total energy of 2,828.30 N-mm. The fabricated functional prototypes are suitable for strain measurement under repetitive tensile forces and maintain greater flexibility, durability, and longevity, making them ideal for intended applications. Also, the dielectric constant of the selected samples was ~2.63, with a simulated specific absorption rate of 0.22 W/kg, indicating consistent, efficient signal transfer with minimal energy loss, making them a preferable choice for real-time monitoring of muscle movement. Further, the results are supported by scanning electron microscopy, energy-dispersive spectroscopy, porosity %, and average grain size.

Keywords: Material extrusion, 3D printing, piezoelectric, functional prototypes, TPU, muscle movement monitoring

3.1 Introduction

Polymer, metal, ceramic, and composite components can be embedded in a build using multimaterial 3D printing (MM3DP) via material extrusion (MEX). This novel additive manufacturing technique significantly enhances the mechanical, thermal, and

***Corresponding author: Gurwinder Singh**, Mechanical Engineering Department, National Institute of Technical Teachers Training and Research, Chandigarh 160019, Chandigarh, India, e-mail: gurwinder14899@gmail.com
Rupinder Singh, Sourav Kumar, Mechanical Engineering Department, National Institute of Technical Teachers Training and Research, Chandigarh, India

 | https://doi.org/10.1515/9783112218846-003

functional properties of various parts [1]. It circumvents the limitations of traditional single MEX by advancing three fundamental techniques in advanced rapid prototyping: delivering precise, controlled deposition of a wide range of materials using innovative techniques, such as multinozzle systems, single-nozzle material switching, and in situ material mixing [2]. Single-nozzle systems reduce hardware complexity but introduce issues such as switching latency and material mixing [3]. In contrast to multinozzle systems, which allow more accurate material placement, a fundamental challenge remains regarding nozzle alignment and contamination. In situ material mixing has enabled the production of functionally graded materials (FGMs) with continuous property gradients [4]. Significant advancements in the mechanical performance of fiber-reinforced MEX have been reported. Short carbon fiber reinforcement increases properties to an optimal concentration (beyond which misalignment tends to reduce performance) [5], whereas continuous carbon fiber reinforcement results in large increases in elastic modulus and strength [6]. There is also full dependence on fiber orientation, i.e., oriented fibers have much more tensile strength than the alternate orientations of fibers [7]. Moreover, the incorporation of conductive fibers and fillers has enabled the development of multipurpose components, including wearable electronics and sensors [8]. Metal-polymer composite filaments, along with debinding and sintering processes, have enabled low-cost manufacturing of compact metallic components in metal extrusion [9]; multimaterial metal printing has enabled the customization of bimetallic structures [10]. Another study highlights that the use of stainless steel has demonstrated enhanced interfacial bonding and improved control over shrinkage behavior after sintering, and thus can be applied to heat exchangers, electrical contacts, and biomedical implants [11, 12]. It is still challenging to control the gradient accurately owing to nonuniform deposition. Still, the MEX process enables the manufacture of FGM with continuously varying material properties, thereby minimizing stress concentrations [13, 14]. Interfacial bonding problems remain in multimaterial polymer extrusion due to thermal and rheological property differences. Still, thermoplastics such as PLA (polylactic acid), ABS (acrylonitrile butadiene styrene), and polyethylene terephthalate glycol can be coupled to achieve user-defined mechanical behavior [15]. The development of polymeric materials reinforced with graphene, referred to as novel smart materials, has enabled their potential use in electronic components and 4D printing [16]. In metal MEX, high sintering temperatures and oxidation are overcome by adopting the new sintering techniques and controlled process conditions, such as an inert gas environment, resulting in better part quality, reduced thermal stresses, and fabricated parts with tunable properties [17]. While the challenges of particle agglomeration and settling remain, ceramics-based MEX has been used to manufacture high-temperature, biocompatible structures from ceramic-polymer composites [18]. After achieving uniform dispersion, enhanced electrical and thermal conductivity in composite materials has been observed using nanofillers such as graphene and carbon nanotubes [19]. Computer-based simulation, such as artificial intelligence (AI), has also played significant roles in these areas by enabling the estimation of macroscopic properties, the adjustment of process parameters, and the optimization of overall production resources in MM3DP systems [20]. Additive manufacturing, pioneered by Hull in the 1980s and paving the way for the layer-by-layer production of complex geometries from digital models, has evolved significantly [21]. In thermoplastics such as ABS, glass fiber reinforcement has been shown to limit shrinkage and enhance rigidity, but at the expense of reduced flexibility. Research beyond this point has mainly focused on improving the material's printability and inline performance [22]. On the other hand, when graphene is added at higher concentrations, it increases mechanical strength but also makes polymer processing difficult due to aggregation [23]. Glycerol and ionic liquids were used as plasticizers to increase viscosity and lower the melting temperature of the organic polymers, thereby facilitating smoother extrusion and addressing printability issues with PVA [24]. These surface-segregating additives have been shown to promote polyglycerol chain entanglement and to impart isotropic mechanical properties, thereby enhancing interlayer adhesion [25]. Although styrene-ethylene-butadiene-styrene and ultra-high-molecular-weight polyethylene blends exhibited improved toughness and reduced stiffness, polymer blending techniques enhanced melt flow behavior and dimensional accuracy when combined with PMMA, as in ABS [24, 26]. Similarly, introducing ABS into poly(2-vinylpyridine) facilitated the extrusion of otherwise brittle polymers, and mixing PBS with PLA significantly improved melt strength, thereby reducing distortion during printing [27, 28]. Molecular design strategies, such as side-chain engineering, have proved that longer chains act as internal plasticizers, resulting in lower viscosity and improved mobility, but at the expense of shape retention [29]. In contrast, higher molecular weight increases resistance to degradation during printing [30]. Moreover, various approaches to chemical modification, e.g., the incorporation of π–π interactions into PET, lead to reduced crystallization and improved adhesion between layers [31], or to hydrogen-bonding networks that result in tunable viscosity and easy processing for polymers such as polyisobutylene (PIB) [32]. Aside from thermoplastics, thermoset materials have also been adapted for extrusion printing. Although cyclotrimerization reactions lead to excellent thermal stability of cyanate ester resins up to 400 °C [33], vitrimers – epoxy systems modified with nano clay that allowed extrusion and showed structural integrity through a controlled curing process [34]. Reversible Diels-Alder chemistry allows thermoset polymers to have tunable viscosity and improved printability [35]. Nano-clay-modified epoxy inks exhibited shear-thinning behavior and excellent structural fidelity upon deposition via direct ink writing [36]. Hydrogels possess tunable rheological properties, making them fascinating materials; for example, thermo-responsive triblock copolymer hydrogels exhibit shear-thinning and fast recovery, promoting stable 3D architectures [37]. In multilayer printing, dual crosslinking strategies, for example, both covalent and physical networks, significantly enhanced hydrogel resilience and prevented structural collapse [38]. Moreover, double-network hydrogels not only benefited from covalent crosslinks and ionic mesh to achieve outstanding mechanical performance (mainly energy dissipation and stretchability) but also showed great promise for high-end applications [39].

The literature indicates that numerous studies have reported the use of functional prototypes in various engineering applications fabricated via the MEX process. However, little has been reported on the 3D printing of piezoelectric functional prototypes using MEX for biomedical applications. The study highlights the fabrication of tensile specimens (ASTM D638 Type V) of thermoplastic polyurethane (TPU) on a cotton-75% lyca-25% material to optimize the fused filament fabrication (FFF)/MEX process for monitoring muscle movement under tensile loading conditions.

3.2 Materials and methods

For the bibliometric analysis, the SCOPUS database over the past 20 years was used to identify research gaps and trends using the keyword "3D printed piezoelectric polymer in wearable sensor," and a total of 20 documents were found and extracted in a . csv format file. Further, an open-source VOS viewer was used to construct and explore bibliometric networks with a minimum of 03 occurrences of terms; of the 802 terms, 51 met the threshold derived from the .csv-extracted file. Additionally, the 51 terms were selected based on their occurrence scores and used to highlight the network diagram. Figure 3.1 presents the network diagram based on the specific keyword "3D printed piezoelectric polymer in wearable sensor." Based on Figure 3.1, Figure 3.2 highlights the role of 3D printing in piezoelectric materials for wearable sensor and device development in biomedical and energy-harvesting applications. Additionally, based on Figure 3.2, Figure 3.3 highlights the research gap: limited studies have been conducted on material-extrusion-based 3D printing for the manufacturing of piezoelectric functional prototypes. This study highlights the fabrication of a 3D-printed thermoplastic TPU polymer and a cotton-lycra-based wearable functional prototype sensor for muscle memory applications.

Figure 3.4 presents country-level data on the 3D-printed piezoelectric polymer used in a wearable sensor. It was noticed that China has the largest number of documents. India contributes a fairly small amount, and Singapore, Israel, and some unknown suppliers also produce modest quantities. Other countries include Nigeria, Germany, Malaysia, the Netherlands, and Australia, among the lowest contributors. In general, the graph shows that China is a key player in this field, and most research is centered in a few leading countries.

In this study, a TPU thermoplastic polymer filament spool and a cotton (75%)-lycra (25%) fabric/cloth were procured from the local market and used to fabricate a functional 3D-printed wearable sensor prototype for muscle memory applications. TPU was selected for its exceptional mechanical properties, elasticity, durability, light-

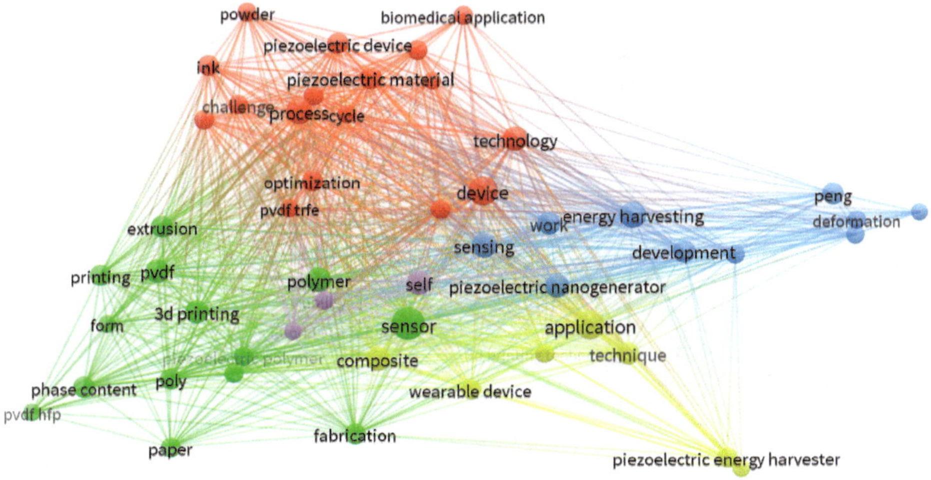

Figure 3.1: Network diagram based on the specific keyword "3D printed piezoelectric polymer in wearable sensor."

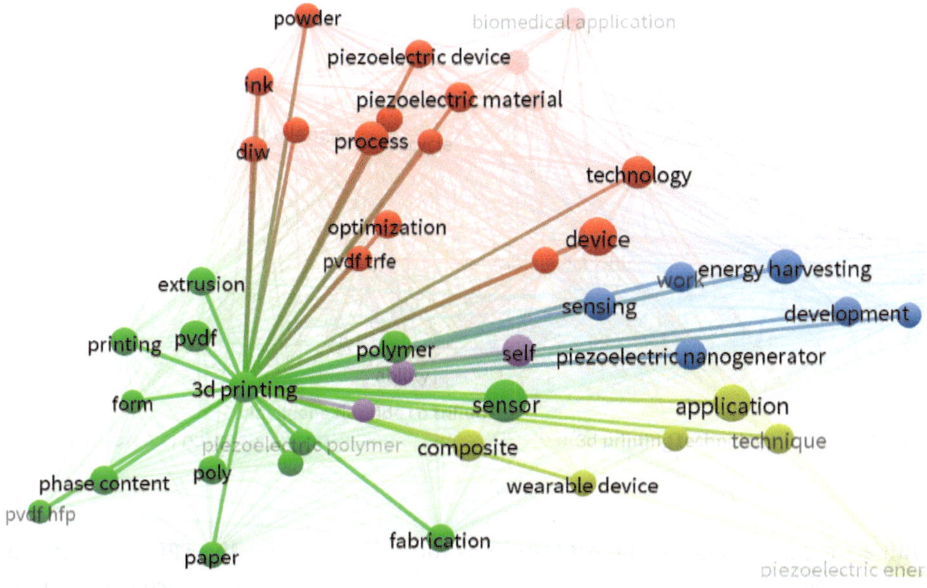

Figure 3.2: Role of 3D printing in functional development systems, such as a piezoelectric sensor and a wearable device.

weight nature, biocompatibility, and comfort, as well as its role as a flexible substrate that provides comfort during body movements. The TPU filament was shredded, and the MFI was measured (as per ASTM D1238) at an extrusion temperature of 230 °C,

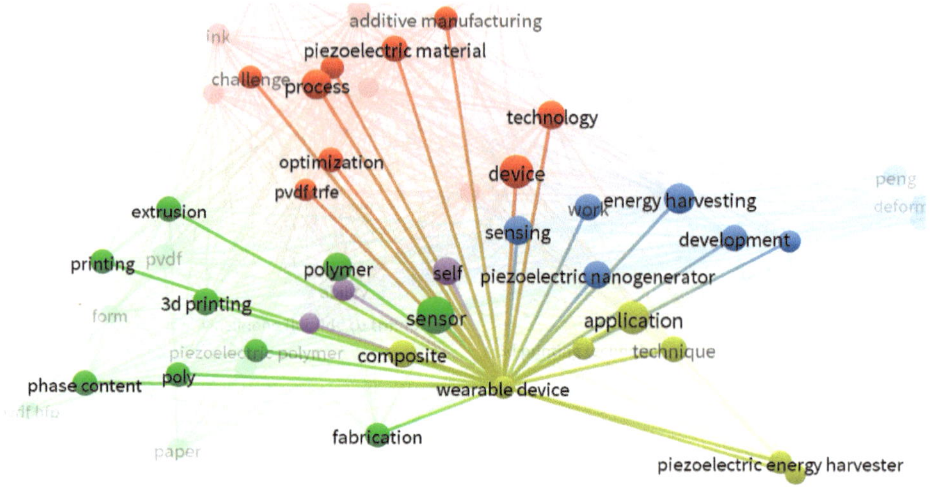

Figure 3.3: Research gap.

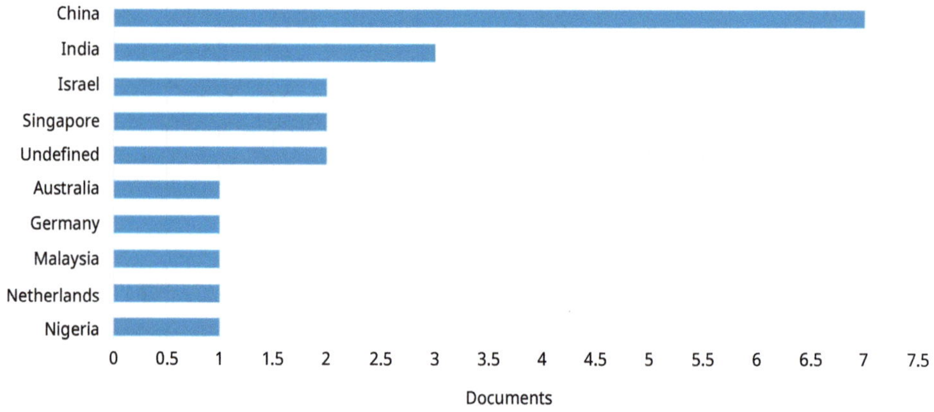

Figure 3.4: Country-wise data reported on the 3D-printed piezoelectric polymer used in a wearable sensor.

with a weight of 2.165 kg. Table 3.1 presents the MFI results for the TPU filament. Further, a design of experiments (DOE) approach using the Taguchi L_9 orthogonal array (OA) was employed to optimize the FFF process. The experiment considered three factors: strain (pre), orientation (°), and infill density (%), having three levels, based on the results from trial runs. Table 3.2 outlines the levels and factors for optimizing the FFF/MEX process. Figure 3.5 shows the adopted methodology used in this study.

Table 3.1: MFI outcomes of TPU polymer.

Iterations	MFI (g/min)
1	4.14
2	3.43
3	3.29
Average	3.62

Table 3.2: Designed DOE based on the three levels and factors.

S. no.	Input parameters		
	Strain (mm)	Orientation (°)	Infill density (%)
1	0	0	60
2	0	45	80
3	0	90	100
4	10	0	80
5	10	45	100
6	10	90	60
7	20	0	100
8	20	45	60
9	20	90	80

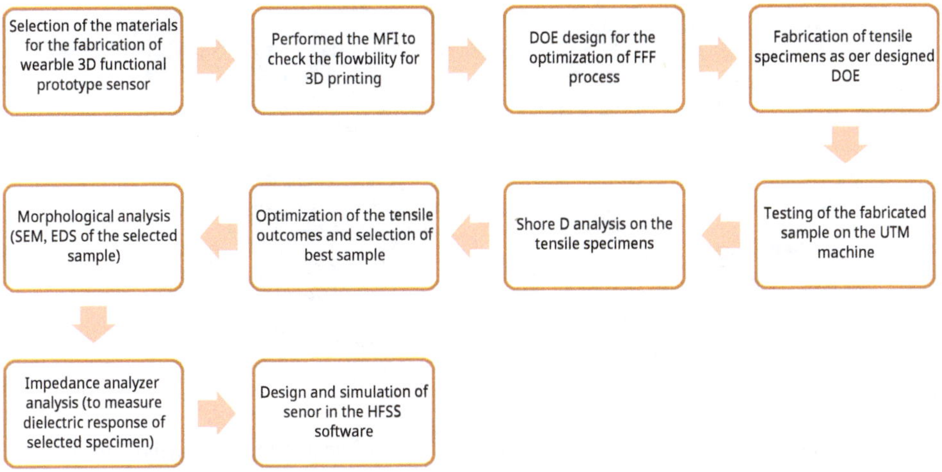

Figure 3.5: Adopted methodology.

The tensile specimens (ASTM-D638 type V) were fabricated from TPU filament and cotton-lycra cloth at different strains (Table 3.1) using an open-source FFF printer with printing parameters based on the designed DOE. Figure 3.6a and b shows the 3D view and the fabricated tensile specimens, along with the cotton-lycra cloth. The tensile test was performed using a universal tensile tester with a maximum capacity of 20 kN and a feed rate of 10 mm/min. To assess the hardness, durability, and rigidity of the fabricated specimens, Shore D hardness testing was performed. Figure 3.7a and b shows the testing of the tensile specimen on the UTM machine and the Shore D analysis of the tensile specimens. The scanning electron microscopy (SEM), energy-dispersive spectroscopy (EDS), porosity %, and average grain size were measured on the selected fractured tensile specimen.

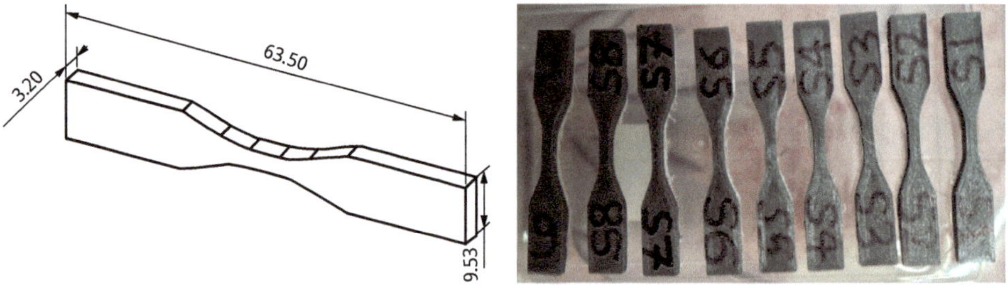

Figure 3.6: (a) 3D view and (b) the fabricated tensile specimens, along with cotton-lycra cloth.

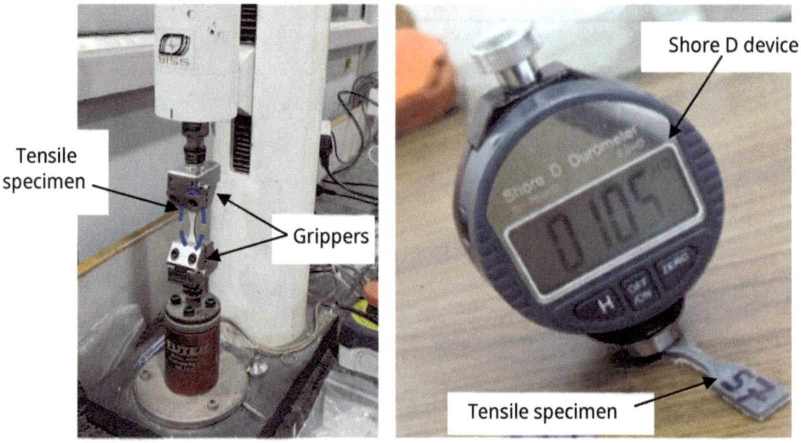

Figure 3.7: (a) Testing of a tensile specimen on the UTM machine and (b) Shore D analysis on the tensile specimen.

3.3 Results and discussion

Table 3.3 shows the observed mechanical test outcomes for the tensile specimens, along with the signal-to-noise (*S*/*N*) ratio. Toughness is the property of a material that absorbs distributed energy before making a fracture, which is also represented as total energy (N-mm) in the tensile specimen. A high total energy indicates that the material possesses enhanced ductility, structural soundness, and resistance to failure under deformation, making it favorable for absorbing high strain energy across different loading environments. According to Table 3.3, Sample S7 has the highest total energy and could therefore be considered the best performer because of its good resistance to repeated deformation, which allows for more flexibility, durability, and longevity; it is ideal for applications such as wearable sensors. Such materials with this property are suitable for strain measurement under repetitive tensile forces while maintaining electrical conductivity, enabling real-time monitoring in muscle memory applications. Tables 3.4 and 3.5 show the ANOVA and response and rank for the *S*/*N* ratio of the observed mechanical property. From Table 3.4, the null probability (*p*) value for the input parameter was found to be significant. It means the selected input parameters do not have much importance on the selected specific output. Figure 3.8 shows the main effect and residual plots of the *S*/*N* ratio of the specific mechanical property. From Figure 3.8, it was observed that, for optimal sample performance in terms of total energy, the suggested parameters were 20 mm strain (pre), 90° orientation, and 80% infill density.

Table 3.3: Experimental outcomes based on the input, output, and *S*/*N* ratio.

S. no.	Input parameters			Output	*S*/*N* ratio	Shore D
	Strain (mm)	Orientation (°)	Infill density (%)	Total energy (N-mm)	Total energy	HD
1	0	0	60	2,372.85	67.51	11.87
2	0	45	80	2,654.39	68.48	15.25
3	0	90	100	2,545.60	68.12	15.12
4	10	0	80	2,638.36	68.43	15.87
5	10	45	100	2,671.34	68.53	16.62
6	10	90	60	2,608.78	68.33	13.87
7	20	0	100	2,828.30	69.03	15.25
8	20	45	60	2,563.75	68.18	13.50
9	20	90	80	2,813.40	69.14	17.37

For the fractured analysis, the SEM was operated at 100× magnification and 5 kV, with an 8-mm working distance. Further, the metallurgical image analysis software was used to measure the porosity (%) and the average grain size of the S7 specimen. Figure 3.9 shows the SEM, EDS, porosity %, and average grain size number of the S7 fractured tensile speci-

Table 3.4: Analysis of variance (ANOVA) for *S/N* ratio.

Observed mechanical properties	Source	DF	Seq SS	Adj SS	Adj MS	F	P
Total energy	Strain	2	0.841	0.841	0.420	3.99	0.200
	Orientation	2	0.065	0.065	0.032	0.31	0.763
	Infill density	2	0.782	0.782	0.391	3.71	0.212
	Residual error	2	0.210	0.210	0.105		
	Total	8	1.900				

Note: DF, degree of freedom; Seq SS, sequential sum of squares; Adj SS, adjusted sum of squares; Adj MS, adjusted mean of squares; *F*, Fisher's value; *P*, probability.

Table 3.5: Response and rank for *S/N* ratio.

Output	Level	Strain	Orientation	Infill density
Total energy	1	68.03	68.32	68
	2	68.43	68.4	68.68
	3	68.78	68.53	68.56
	Delta	0.75	0.21	0.68
	Rank	1	3	2

men. From Figure 3.9, it was observed that the S7 sample shows a porosity of 27.53% and an average grain size of 5.25. The S7 sample with a porosity of 27.53% and an average grain size of 5.25 shows many voids in its structure. This level of porosity may compromise mechanical behavior and, under tensile loading, lead to premature deformation or failure by decreasing the effective load-bearing cross-sectional area and creating stress concentration sites. However, such porosity may also be beneficial in wearable sensor and muscle memory applications, making it more flexible, less stiff, and more compressible under dynamic conditions. The material's high total energy demonstrates its porous structure, which allows greater surface conformability to body dynamics and repeatable deformation while maintaining sufficient energy-absorbing capability.

The impedance analyzer was used (1 Hz to 8 MHz) on the circular disk of ϕ10 mm with 1.7 mm thickness to measure the relative dielectric constant (ε) of the selected tensile specimen. The Debye relaxation equations (3.1) and (3.2) were used for the prediction of the dielectric constant (ε') and loss factor (ε'') at the industrial, scientific, and medical (ISM) band. Table 3.6 shows the observed sensing outcomes of the S7 sample. The observed ε' of 2.63 indicates that it may store electrical energy in an electric field, making it suitable for wearable sensors and muscle memory applications and enabling consistent, efficient signal transfer with minimal energy loss, which enables reliable sensing applications that require electrical performance, as well as increased mobility:

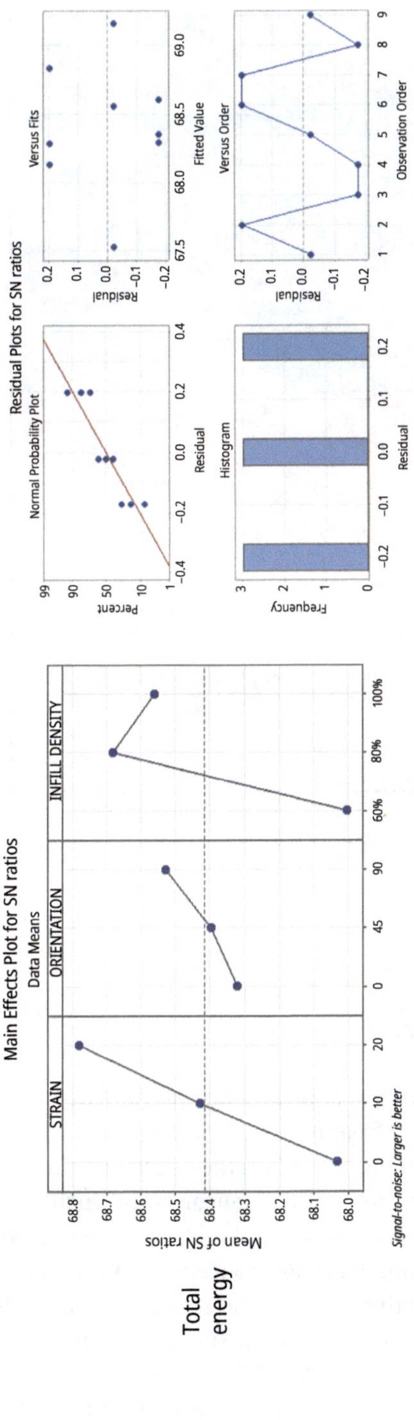

Figure 3.8: Main effects and residual plots of observed mechanical properties.

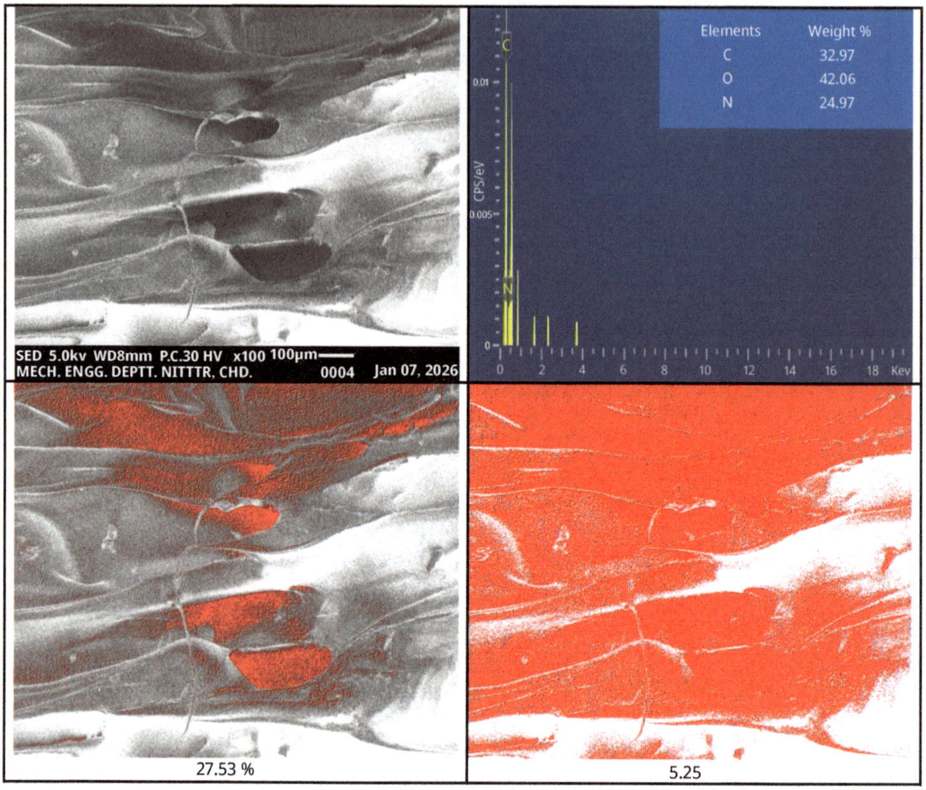

Figure 3.9: SEM, EDS, porosity %, and average grain size number of S7 specimen.

$$\in' = \in_\infty + \frac{\in_S - \in_\infty}{1 + (\omega\tau)^2} \tag{3.1}$$

$$\in'' = \frac{(\in_S - \in_\infty)\omega\tau}{1 + (\omega\tau)^2} \tag{3.2}$$

where ε' is the dielectric constant, ε'' is the loss factor, ε_s is the static permittivity, ε_∞ is the highest dielectric permittivity of the required frequency (taking 2.57), ω is the angular frequency of the applied field, and τ (assuming 1×10^{-9}) is the relaxation time.

Based on the observed ε' and ε'', the dimensions of the sensor were calculated (eqs. (3.3)–(3.8)) and simulated using the high-frequency simulator software (HFSS). The sensor was based on the microstrip patch antenna (MPA) concept. MPA comprises a nonconductive part, the substrate, and conductive (ground and patch) layers, much like a sandwich. Generally, Cu alloy is used to make the conducting parts. For the S7 sample sensor, the substrate dimensions were 47.20 × 55.50 × 1.7 mm, and for the conductive patch and ground, the dimensions were 45.50 × 37 × 0.08 mm and 47.20 × 55.50 × 0.08 mm, respectively. Figure 3.10 shows the sensor design. Figure 3.11 shows

Table 3.6: Observed sensing properties of the S7 sample.

Sample	Observations		
	Observed ε_r at 8 MHz	Predicted dielectric constant using the Debye relaxation model	
		ε'	ε''
S7	34.41	2.63	0.078

the reflection coefficient S_{11} versus R_f of the desired sensor. The simulated sensor's return loss at 2.40 GHz is −13 dB, indicating good signal reflection and impedance matching, which indeed guarantee stable sensor operation. The sensor's resonance behavior suggests its suitability for wearables, wireless, battery-free, and muscle-memory applications in the ISM band. The observed specific absorption rate (SAR) (Figure 3.12) is within the permissible and acceptable range, poses no harmful thermal effects, and is suitable for the intended application. The following eqs. (3.3)–(3.8) were used for calculations of dimensions of MPA based on the reported literature [40].

The width (w) of the patch is given by

$$w = \frac{c}{2f_r}\sqrt{2/(\varepsilon_r+1)} \tag{3.3}$$

To determine the fringe factor (ΔL), the effective dielectric constant (ΔL) (ε_{eff}) needs to be calculated, i.e.,

$$\varepsilon_{\text{eff}} = \left[\frac{(\varepsilon_r+1)}{2}\right] + \left[\left(\frac{(\varepsilon_r-1)}{2}\right)\left(1+12\frac{h}{w}\right)^{-0.5}\right] \tag{3.4}$$

$$\Delta L = 0.42h\left[\frac{(\varepsilon_{\text{eff}}+0.3)\left(\frac{w}{h}+0.264\right)}{(\varepsilon_{\text{eff}}-0.258)\left(\frac{w}{h}+0.8\right)}\right] \tag{3.5}$$

The length (L) of the patch is given by

$$L = \frac{1}{2f_r\sqrt{(\varepsilon_{\text{eff}}\varepsilon_o\mu_o)}} - 2\Delta L \tag{3.6}$$

The dimensions of the substrate were given by Eqs. 3.7 and Eqs. 3.8:

$$W_s = W + 6 \tag{3.7}$$

$$L_s = L + 6h \tag{3.8}$$

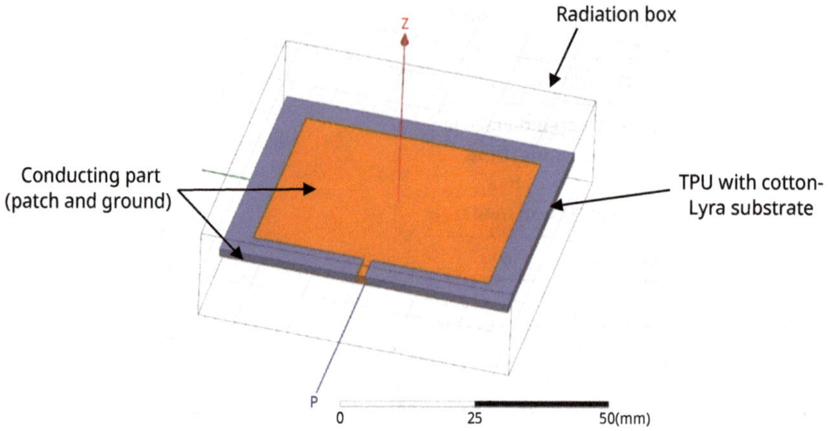

Figure 3.10: Sensor design.

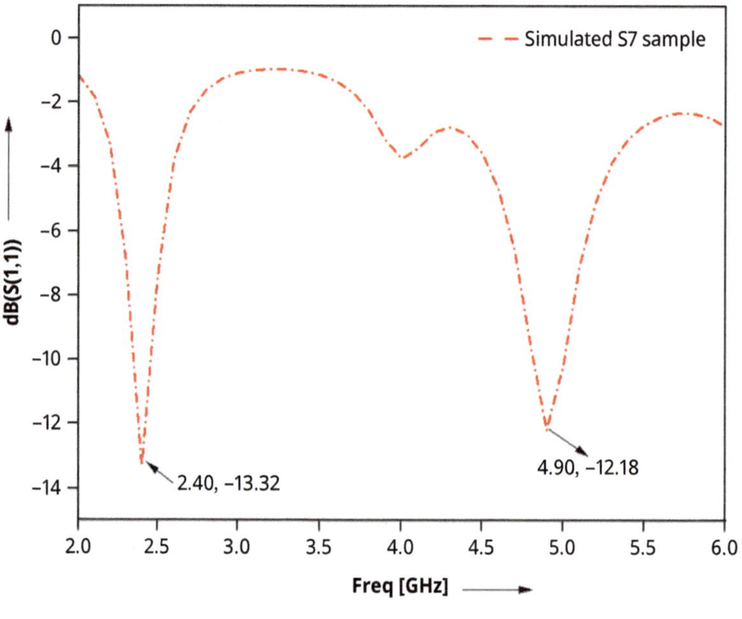

Figure 3.11: S_{11} versus frequency of S7 sample.

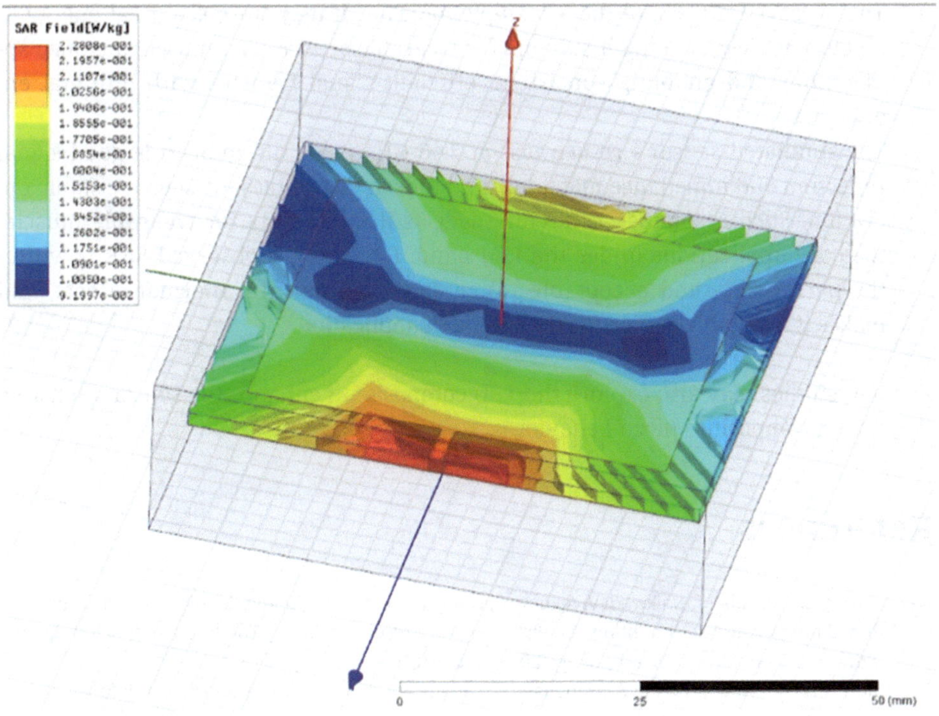

Figure 3.12: Simulated SAR of S7 sample.

3.4 Conclusions

This chapter highlights the fabrication of tensile specimens of TPU with cotton (75%)-lycra (25%) using the 3D-printed MEX process based on the DOE for real-time monitoring of muscle memory applications. The following are the conclusions:

- The S7 sample showed the highest total energy, making it the best performer for the intended applications. Such materials with this property are suitable for strain measurement under repetitive tensile forces while maintaining electrical conductivity, enabling real-time monitoring in muscle memory applications.
- Based on the ANOVA observation, the null probability (p) value for the input parameter was found to be significant. It means the selected input parameters do not have much importance on the selected specific output.
- The S7 sample showed a porosity of 27.53% and an average grain size of 5.25, indicating that it exhibits many voids in its structure. This level of porosity may compromise mechanical behavior under tensile loading. However, such porosity may also be beneficial in wearable sensor and muscle memory applications, making it more flexible, less stiff, and more compressible under dynamic conditions.

- The observed ε' (2.63) of the S7 sample indicates it may store electrical energy in an electric field, making it suitable for wearable sensors and muscle memory applications and enabling consistent, efficient signal transfer with minimal energy loss.
- The simulated sensor's return loss at 2.40 GHz is −13 dB, indicating good signal reflection and impedance matching, and indeed guarantees the sensor's stable operation. Such resonance behavior suggests its suitability for wearable muscle-memory applications under the ISM band. Further, the observed SAR is within the permissible and acceptable range, poses no harmful thermal effects, and makes it a suitable choice for the intended application.

Further, studies may also explore the TPU composite to enhance the sensor's sensitivity and the composite's piezo-behavior.

References

[1] Rajeshirke M, Fidan I, Naikwadi V, Alkunte S, Gupta A, Mohammadizadeh M. Material extrusion-based multi-material 3D printing: a holistic review of recent advances. The International Journal of Advanced Manufacturing Technology. 2025 Jul;139(1):149–74.

[2] Rajeshirke M, Fidan I, Gupta A, Mäntyjärvi K. Fatigue analysis of short carbon fiber reinforced composite components manufactured using fiber-reinforced additive manufacturing. https://doi.org/10.26153/TSW/44139

[3] Rajeshirke M, Alkunte S, Huseynov O, Fidan I. Fatigue analysis of additively manufactured short carbon fiber-reinforced PETG Components. The International Journal of Advanced Manufacturing Technology. 2023 Sep;128(5):2377–94.

[4] Alkunte S, Rajeshirke M, Fidan I, Hasanov S. Performance evaluation of fatigue behavior in extrusion-based functionally graded materials. The International Journal of Advanced Manufacturing Technology. 2023 Sep;128(1):863–75.

[5] Ahmed A, Arya S, Gupta V, Furukawa H, Khosla A. 4D printing: Fundamentals, materials, applications and challenges. Polymer. 2021 Jul 16;228:123926.

[6] Melenka GW, Cheung BK, Schofield JS, Dawson MR, Carey JP. Evaluation and prediction of the tensile properties of continuous fiber-reinforced 3D printed structures. Composite Structures. 2016 Oct 1;153:866–75.

[7] Papa I, Silvestri AT, Ricciardi MR, Lopresto V, Squillace A. Effect of fibre orientation on novel continuous 3D-printed fiber-reinforced composites. Polymers. 2021 Jul 30;13(15):2524.

[8] Nag A, Nuthalapati S, Mukhopadhyay SC. Carbon fiber/polymer-based composites for wearable sensors: A review. IEEE Sensors Journal. 2022 Apr 25;22(11):10235–45.

[9] Mazeeva A, Masaylo D, Konov G, Popovich A. Multi-metal additive manufacturing by extrusion-based 3D printing for structural applications: A review. Metals. 2024 Nov 16;14(11):1296.

[10] Nazir A, Gokcekaya O, Billah KM, Ertugrul O, Jiang J, Sun J, Hussain S. Multi-material additive manufacturing: A systematic review of design, properties, applications, challenges, and 3D printing of materials and cellular metamaterials. Materials & Design. 2023 Feb 1;226:111661.

[11] Asami K, Lozares JM, Ullah A, Bossen B, Clague L, Emmelmann C. Material extrusion of metals: Enabling multi-material alloys in additive manufacturing. Materials Today Communications. 2024 Mar 1;38:107889.

[12] Cheng J, Wan L, Cai Y, Zhu J, Song P, Dong J. Fabrication of W–20 wt.% Cu alloys by powder injection molding. Journal of Materials Processing Technology. 2010 Jan 1;210(1):137–42.
[13] Hasanov S, Gupta A, Alifui-Segbaya F, Fidan I. Hierarchical homogenization and experimental evaluation of functionally graded materials manufactured by the fused filament fabrication process. Composite Structures. 2021 Nov 1;275:114488.
[14] You S, Jiang D, Wang F, Ning F. Anisotropic sintering shrinkage behavior of stainless steel fabricated by extrusion-based metal additive manufacturing. Journal of Manufacturing Processes. 2023 Sep 8;101:1508–20.
[15] Baca D, Ahmad R. The impact on the mechanical properties of multi-material polymers fabricated with a single mixing nozzle and multi-nozzle systems via fused deposition modeling. The International Journal of Advanced Manufacturing Technology. 2020 Feb;106(9):4509–20.
[16] Noh JS. Conductive elastomers for stretchable electronics, sensors, and energy harvesters. Polymers. 2016 Apr 5;8(4):123.
[17] Yue W, Zhang Y, Zheng Z, Lai Y. Hybrid laser additive manufacturing of metals: A review. Coatings. 2024 Mar 6;14(3):315.
[18] Wick-Joliat R, Schroffenegger M, Penner D. Multi-material ceramic material extrusion 3D printing with granulated injection molding feedstocks. Ceramics International. 2023 Feb 15;49(4):6361–67.
[19] Podsiadły B, Matuszewski P, Skalski A, Słoma M. Carbon nanotube-based composite filaments for 3d printing of structural and conductive elements. Applied Sciences. 2021 Jan 30;11(3):1272.
[20] Bhat C, Prajapati MJ, Kumar A, Jeng JY. Additive manufacturing-enabled advanced design and process strategies for multi-functional lattice structures. Materials. 2024 Jul 9;17(14):3398.
[21] Hull CW, Gabriel S. Apparatus for production of three-dimensional objects by stereolithography US Patent 4575330. fabbstercom. 1986.
[22] Zhong W, Li F, Zhang Z, Song L, Li Z. Short fiber reinforced composites for fused deposition modeling. Materials Science and Engineering: A. 2001 Mar 31;301(2):125–30.
[23] Wei X, Li D, Jiang W, Gu Z, Wang X, Zhang Z, Sun Z. 3D printable graphene composite. Scientific Reports. 2015;5(1):11181. [Internet].
[24] Chen S, Lu J, Feng J. 3D-Printable ABS blends with improved scratch resistance and balanced mechanical performance. Industrial & Engineering Chemistry Research. 2018b Mar 2;57(11):3923–31.
[25] Levenhagen NP, Dadmun MD. Reactive processing in extrusion-based 3D printing to improve isotropy and mechanical properties. Macromolecules. 2019 Aug 22;52(17):6495–501.
[26] Rocha CR, Perez AR, Roberson DA, Shemelya CM, MacDonald E, Wicker RB. Novel ABS-based binary and ternary polymer blends for material extrusion 3D printing. Journal of Materials Research. 2014 Sep;29(17):1859–66.
[27] Nadgorny M, Xiao Z, Chen C, Connal LA. Three-dimensional printing of pH-responsive and functional polymers on an affordable desktop printer. ACS applied materials & interfaces. 2016 Oct 26;8 (42):28946–54.
[28] Ou-Yang Q, Guo B, Xu J. Preparation and characterization of poly (butylene succinate)/polylactide blends for fused deposition modeling 3D printing. ACS omega. 2018 Oct 29;3(10):14309–17.
[29] Jain S, Fuoco T, Yassin MA, Mustafa K, Finne-Wistrand A. Printability and critical insight into polymer properties during direct-extrusion based 3D printing of medical grade polylactide and copolyesters. Biomacromolecules. 2019a Sep 30;21(2):388–96.
[30] Jain T, Clay W, Tseng YM, Vishwakarma A, Narayanan A, Ortiz D, Liu Q, Joy A. Role of pendant side-chain length in determining polymer 3D printability. Polymer Chemistry. 2019b;10(40):5543–54.
[31] Chen L, Zhao HB, Ni YP, Fu T, Wu WS, Wang XL, Wang YZ. 3D printable robust shape memory PET copolyesters with fire safety via π-stacking and synergistic crosslinking. Journal of Materials Chemistry A. 2019;7(28):17037–45.

[32] Rupp H, Döhler D, Hilgeroth P, Mahmood N, Beiner M, Binder WH. 3D printing of supramolecular polymers: Impact of nanoparticles and phase separation on printability. Macromolecular Rapid Communications. 2019 Dec;40(24):1900467.

[33] Shi Q, Yu K, Kuang X, Mu X, Dunn CK, Dunn ML, Wang T, Qi HJ. Recyclable 3D printing of vitrimer epoxy. Materials Horizons. 2017;4(4):598–607.

[34] Chandrasekaran S, Duoss EB, Worsley MA, Lewicki JP. 3D printing of high-performance cyanate ester thermoset polymers. Journal of Materials Chemistry A. 2018;6(3):853–58.

[35] Chen G, Chen N, Wang Q. Preparation of poly (vinyl alcohol)/ionic liquid composites with improved processability and electrical conductivity for fused deposition modeling. Materials & design. 2018a Nov 5;157:273–83.

[36] Lewicki JP, Rodriguez JN, Zhu C, Worsley MA, Wu AS, Kanarska Y, Horn JD, Duoss EB, Ortega JM, Elmer W, Hensleigh R. 3D-printing of meso-structurally ordered carbon fiber/polymer composites with unprecedented orthotropic physical properties. Scientific reports. 2017 Mar 6 7(1):43401.

[37] Zhang M, Vora A, Han W, Wojtecki RJ, Maune H, Le AB, Thompson LE, McClelland GM, Ribet F, Engler AC, Nelson A. Dual-responsive hydrogels for direct-write 3D printing. Macromolecules. 2015 Sep 22;48(18):6482–88.

[38] Highley CB, Rodell CB, Burdick JA. Direct 3D Printing of Shear-Thinning Hydrogels into Self-Healing Hydrogels. Advanced Materials (Deerfield Beach, Fla.). 2015 Jul 15;27(34):5075–79.

[39] Hong S, Sycks D, Chan HF, Lin S, Lopez GP, Guilak F, Leong KW, Zhao X. 3D printing of highly stretchable and tough hydrogels into complex, cellularized structures. Advanced materials (Deerfield Beach, Fla.). 2015 Jun 1;27(27):4035.

[40] Memon MI, Paliwal A. Microstrip patch antenna design calculator. International Journal of Engineering Research & Technology IJERT. 2013 Sep;2:3080–84.

Minhaz Husain*, Rupinder Singh, Gurwinder Singh

Chapter 4
Stereolithography and digital light processing-based 3D printing for piezoelectric materials

Abstract: This chapter provides a detailed overview of stereolithography (SLA) and digital light processing (DLP) additive manufacturing routes for the fabrication of piezoelectric materials, covering materials, processing principles, and structure-property relationships. The vat photopolymerization of piezoelectric polymers, ceramics, and composite systems is considered with respect to their formulation requirements, curing characteristics, and post-curing procedures. The basic principles of photopolymerization, along with their effects on curing depth, resolution, and interlayer bonding, are described, followed by a closer look at SLA and DLP procurement methods for the production of three-dimensional piezoelectric actuators. Special emphasis is placed on ceramic-filled photopolymer resins, in which light scattering, viscosity, and solid loading control both printability and the final electromechanical response. Key process parameters, such as exposure energy, layer thickness, particle size, and build orientation, are investigated for both processes, and their impact on dimensional accuracy, microstructure homogeneity, and the piezoelectric coefficient is discussed. Post-processing stages, including cleaning, ultraviolet post-curing, debinding, sintering, and electrical poling, are essential steps for activating and optimizing piezoelectric functionality. Also, compare SLA to DLP and evaluate trade-offs in resolution, print speed, scalability, and material integration, concluding that they don't compete but rather complement each other. The use of piezoelectric materials printed with SLA and DLP in sensors, actuators, energy-harvesting systems, biomedical equipment, microelectromechanical devices, and innovative constructions is also reported.

Keywords: 3D printing, stereolithography, digital light processing, piezoelectric materials

***Corresponding author: Minhaz Husain,** Mechanical Engineering Department, Marwadi University Research Center, Marwadi University, Rajkot, Gujarat, India, e-mail: minhazme2012@gmail.com
Rupinder Singh, Gurwinder Singh, Mechanical Engineering Department, National Institute of Technical Teachers Training and Research, Chandigarh, India

 | https://doi.org/10.1515/9783112218846-004

4.1 Introduction

Piezoelectric materials, which can convert mechanical energy into electrical energy and vice versa, have been widely applied in many engineering fields. This electromechanical coupling has found numerous applications in sensors, actuators, energy harvesters, ultrasonic transducers, structural health monitoring systems, and the recently emerging wearable and implantable devices [1]. Conventionally, piezoelectric devices are produced using subtractive manufacturing processes (such as machining idealized geometric shapes), tape casting, and molding, thereby imposing constraints on acceptable geometry, material waste, and design freedom. With the development of micro- and nanotechnology and integrated functionality, such methods are less able to meet the demands of complex shapes, gradient structures, and tailored designs [2]. The earliest commercially available additive manufacturing (AM) process was stereolithography (SLA), developed in the late 1980s. It is equipped with a pinpoint ultraviolet (UV) LED light source and guided by a UV laser to harden layers of photosensitive resin separately, forming an accurate solid object. Digital light processing (DLP), developed later, projects light into the solidifying liquid photocurable resin to cure it layer by layer, resulting in a marked increase in printing speed and improved in-layer uniformity. These two processes have been widely modified for processing piezoelectric materials by employing functionalized photopolymer composites, ceramic-loaded resins, and post-processing techniques such as debinding and sintering [3]. The SLA and DLP with piezoelectric materials are convergent and have enabled the prototyping of complex electromechanical systems that were previously infeasible or impossible to fabricate [4]. These technologies allow the creation of lattice patterns, porous architectures, graded material patterns, and microscale features, which can lead to promising piezoelectric sensitivity, mechanical compliance, and energy conversion efficiency. The digitalization of processes aligns well with rapid prototyping, customization, and integration in innovative manufacturing ecosystems [5].

Three-dimensional (3D) printing, also known as AM, is a disruptive fabrication method that, in addition to addressing many drawbacks, enables the design of functional devices at the micro- and nanoscales [6]. AM, where parts are built up layer by layer from digital models, allows more control over geometry, internal structure, and material distribution than traditional subtractive processes [7]. Among AM systems, vat photopolymerization processes, such as SLA and DLP, have generally emerged as attractive methods for fabricating high-resolution polymer and ceramic-based structures [3]. Their capabilities in achieving satisfactory feature resolution, polishing to a mate surface finish, and dimensional accuracy are desirable for piezoelectric materials, whose functional performance is highly dependent on microstructural homogeneity and geometric control [8].

This chapter discusses SLA and DLP-based 3D printing methods for fabricating piezoelectric materials. The focus is on process structure property relationships and on comparing different SLA and DLP methods and their potential applications in next-generation piezoelectric devices.

4.2 Vat photopolymerization

A family of AM processes that utilizes light to cure liquid photopolymer resins, typically in the form of additives, enabling the creation of solid parts layer by layer. The process of interest in all these cases is photopolymerization, which converts monomers or oligomers into a network of polymers via a light-activated chemical reaction [9]. For piezoelectric material processing, vat photopolymerization is a versatile tool for building functional composites and ceramic precursors with fine control over shape.

4.2.1 Photopolymerization mechanism

There are three essential ingredients in all photopolymerizations: a photopolymer resin, a photoinitiator, and light of an appropriate wavelength. UV irradiation, the photoinitiator absorbs a photon and produces reactive species, such as free radicals or cations. Such reactive species act as initiators in the polymerization process, attacking monomer double bonds to enable chain growth and cross-linking [10]. The intensity of light determines the degree of curing, the exposure duration, the resin's formulation, and its optical penetration depth. The curing depth (D_c) in vat photopolymerization can be explained by Beer-Lambert's law, as per the following equation [11]:

$$D_c = D_p \ln\left(\frac{E_o}{E_c}\right) \tag{4.1}$$

where D_p is the penetration depth of light in the resin, E_o indicates the applied exposure energy, and E_c is the critical energy required for initiating polymerization.

4.2.2 A piezoelectric material in photopolymer systems

Piezoelectric materials for vat photopolymerization can be roughly divided into polymer-based, ceramic-based, and composite piezoelectric systems [12]. Piezoelectric polymers like polyvinylidene fluoride (PVDF) and its copolymers are favored as a result of their flexible, low-density structure, as well as for their ability to be combined with photopolymer resins [13]. However, their piezoelectric coefficients are lower than those of ceramic materials [14]. Lead zirconate titanate (PZT), barium titanate ($BaTiO_3$)

[15], and potassium sodium niobate (KNN)-based piezoelectric ceramics show a high performance in piezoelectricity [16]. Still, because they are inherently brittle, they must be sintered at high temperatures. In vat photopolymerization, incorporation of these ceramics is typically achieved by dispersing a ceramic powder in a photocurable resin to form a slurry. The printed green body is then debound and sintered to obtain dense ceramic structures. Composite systems of piezoelectric ceramic particles in a polymer matrix offer a compromise between mechanical flexibility and functional performance. Vat photopolymerization enables control of particle content and spatial distribution, as well as internal architecture, which are essential for maximizing electromechanical coupling [17].

4.2.3 Characteristics of the process and their effect

Several process variables influence the quality and performance of vat-photopolymerized piezoelectric devices (Table 4.1) [18]. Thickness affects surface finish, dimensional accuracy, and build time. Curing depth and interlayer adhesion are affected by exposure energy, whereas the resin's viscosity determines particle sedimentation and print stability [19]. For resin filled with ceramic particles, the size distribution of the particles and the particle loading in solids are both critical for light scattering, curing behavior, and final density after sintering.

Table 4.1: Key parameters to influence the structural and functional properties of the vat photopolymerization process.

S. no.	Key parameters	Parameters
1	Material formulation	Photoinitiator concentration, viscosity, fillers, absorber, pigments, particle size, and loading
2	Printing and exposure	Exposure time and energy dosage, layer thickness, light source parameters, part orientation, and support structure
3	Process and post-processing	Post-curing time and intensity, cleaning process
4	Optimized parameter interdependencies	Interlayer adhesion, aspect ratio, and feature size, dimensional accuracy versus speed

4.3 3D printing of piezoelectric materials using SLA

SLA is one of the oldest vat photopolymerization technologies and a reference for high-resolution AM. Thus, its application to piezoelectric materials arises from the ability of electron beam-induced deposition to readily create complex 3D geometric structures with smooth surfaces, precise dimensional control, and high feature resolution, with direct implications for piezoelectric performance [20]. In SLA-based production of piezoelectric devices, the method may also be adjusted to backfill functional polymer and ceramic-loaded photopolymer resins, as well as composite systems that require careful management of curing and microstructural development. A schematic of the SLA process for fabricating 3D skull structures is shown in Figure 4.1, which highlights the main components: a laser source, a laser beam, a build platform and its movement, a photopolymer resin, and a cured object layer by layer, as well as the scanning direction of the laser.

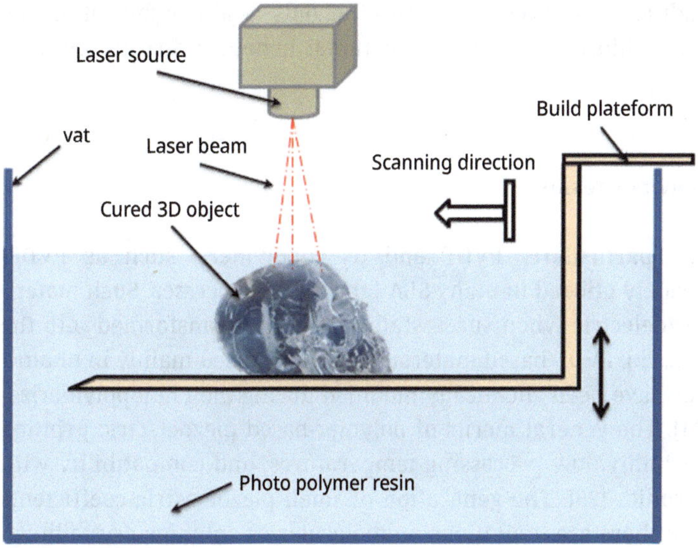

Figure 4.1: Schematic diagram of the vat photopolymerization process.

4.3.1 Working principle of SLA

In SLA, a focused UV laser is directed onto the surface of a vat or onto a horizontal window below the surface of a photocurable liquid resin (Figure 4.1). The laser follows the part's cross-sectional geometry, described by a sliced digital model, curing the resin layer by layer via photopolymerization. When a layer is finished, the build platform lowers by the layer thickness, and uncured resin is recoated over the previous cross section. In contrast, the next layer's exposure brings new resin to the surface [20], which is repeated, and a 3D structure is built up layer by layer. There are two types of SLA configuration available in the market, i.e., top-down and bottom-up. In top-down systems, the laser travels over the free surface of the resin, and the print bed descends into the vat, providing a larger building volume and an easier recoat but consuming a significant amount of resin. The bottom-up approach, which is more common in desktop and laboratory-scale printers, cures layers against a transparent vat window as the build platform rises [21]. Bottom-up systems result in lower resin consumption and provide higher-resolution prints, but they introduce complications related to peel forces and layer separation.

4.3.2 Printing with piezoelectric materials by SLA of various materials

The success of SLA for piezoelectric materials is closely tied to the formulation of the resin. Compared to traditional SLA resins, which offer only aesthetic appeal or mechanical properties, piezo-bioresins provide functional behavior directly or post-process.

4.3.2.1 Piezoelectric polymer resins

Piezoelectric polymers, particularly PVDF and its copolymers, such as PVDF-trifluoroethylene, are widely utilized in many SLA fabrication processes. Such materials are intrinsically piezoelectric when successfully poled and transformed into the electroactive β-phase [22, 23]. PVDF-based materials have been used mainly in photocurable resin matrices or have been chemically modified during the photopolymerization process in SLA [24]. The general merits of polymer-based piezoelectric printing include mechanical flexibility, low processing temperatures, and compatibility with wearable/flexible electronics [25]. The generation of small piezoelectric coefficients has been a problem, and there are limitations on molecular orientation, crystallinity, and electric poling efficiency.

4.3.2.2 Ceramic-filled photopolymer resins

Ceramic-based piezoelectric materials, such as PZT, $BaTiO_3$, and KNN, have much higher piezoelectric coefficients than polymers [26]. They are used in SLA as fine powders dispersed in a photopolymerizable resin to provide a high-loaded slurry. The most typical solid loading of usable ceramics, to achieve a balance between printability and final density, is 40–65 by vol%. Ceramic-filled resin enables the manufacture of complex piezoelectric shapes, which would be challenging to produce using conventional ceramic processing methods. After 3D printing, the green parts are debound to remove the polymeric binder. They are then sintered at high temperatures to densify the ceramic and achieve piezoelectric properties [27]. Light attenuation and absorption by ceramic particles are key to the SLA curing process. It is very challenging to control curing depth and interlaminar adhesion by adjusting laser power, scanning speed, and hatch spacing [28].

4.3.3 Process parameters and optimization of SLA

Several processing parameters influence the quality and functional performance of SLA printed piezoelectric materials. Energy is a function of laser power and speed; thus, exposure energy affects both cure depth and lateral resolution. The hatch distribution defines intersections with adjacent scan lines, which, in turn, affects surface smoothness and uniformity [2]. The thickness of the layers dictates build time, surface quality, and part accuracy. In ceramic-filled systems, curing effectiveness is lower than in neat systems due to light scattering and can be estimated to some extent using an amended Beer-Lambert equation. Such inadequate curing is likely to result in poor interlayer adhesion, and overexposure leads to dimensional errors and loss of feature details. Table 4.2 presents the essential SLA process parameters and their impact on the quality of piezoelectric parts [6].

Table 4.2: Stereolithography process parameters and their effect on SLA-printed piezoelectric materials.

Process parameter	Optimal setting/range	Effect on piezoelectric properties/print quality
Filler content (vol%)	40–80 vol%, which depended on the filler type	High filler (>50 vol%) reduces sintering shrinkage but increases viscosity, requiring a longer exposure time. A high filler content (>70 wt%) can improve the dielectric constant (d_{33}).
Layer thickness	10–100 μm (typically 20–50 μm for ceramics)	Thin layer (10–20 μm) increases print resolution but may cause delamination; 50–100 μm is better for structural integrity.
Exposure time/ intensity	2–16 s (depend on resin/ filler)	Increase polymerization depth – too high causes excessive scattering, too low results in weak interlayer binding.
Filler particle size	Submicron to nano (200–600 nm)	Smaller particles improve green density but cause intense light scattering, reducing cure depth; 500 nm to 1 μm is often preferred.
Build orientation	Vertical or tilted	Affects structural integrity and support requirements. Vertical orientation can improve tensile strength.
Post-curing time	Specific to resin (minutes to hours)	Increases cross-linking, which can reduce green-part brittleness but might make debinding more difficult.
Debinding rate	Very slow (0.1–1 °C/min)	Crucial to avoid cracks in piezoelectric green bodies due to thermal expansion and gas evolution.

4.3.4 Post-processing and piezoelectric activation

Post-processing poses a significant challenge to the manufacturing of piezoelectric materials using SLA. Most printed polymeric and composite parts require cleaning with a solvent to remove uncured resin, followed by post-curing under UV light to complete polymerization. Post-treatments, such as thermal debinding and sintering [29], which are common to ceramic bodies, result in distinct densities and grain sizes, and thus distinct piezoelectric properties [30]. Poling is required to align the ferroelectric domains in piezoelectric ceramics after sintering. This poling is often achieved by heating the system and applying a strong electric field, resulting in permanent dipole alignment. The piezoelectric coefficient (d_{33}) obtained is the primary performance parameter, and its value varies significantly with microstructure, porosity, and domain orientation. Electrical poling is the final step in activating piezoelectric properties in ferroelectric materials. Poling is a process of aligning ferroelectric domains in a single preferred direction by applying a high electric field, typically at an elevated temperature [31]. A field exceeding the material's coercive field must be used to maintain the domain orientation. For porous or composite piezoelectric structures, the field distribution is nonuniform and thus may lead to spatial differences in the piezoelectric response. The poling conditions, therefore, need to be customized to the particular architecture and material system:

$$d_{33} = \frac{D}{\sigma} \tag{4.2}$$

The piezoelectric charge coefficient d_{33}, which refers to the electric displacement produced for a given applied stress, has been generally used as an index of poling efficiency:where D is the electric displacement and σ is the applied mechanical stress eq. (4.2) [32]. When using SLA and DLP 3D printing, the main goal is to achieve high, stable d_{33} values for piezoelectric manufacturing.

4.3.5 Advantages and disadvantages of SLA for piezoelectric materials

Piezoelectric fabrication using SLA is attractive for several reasons, including high surface finish, good geometric fidelity in a correspondence sense, and the ability to print complex structures, among others. These properties facilitate high electromechanical coupling and excellent performance. SLA also has some limitations. Since laser scanning is point-by-point, printing is a low-speed process and would be slower for large cross-sectional areas [33]. Filled resins containing ceramic are challenging to handle due to their high viscosity and poor optical penetration. The long-term material stability and scale-up of these materials are subjects of continued study. SLA remains a powerful technique for constructing high-performance piezoelectric materials, particularly where precision, tailored designs, and complex geometries are required.

4.4 Fabrication of piezoelectric materials using DLP

DLP is one of the more advanced vat photopolymerization processes, which has begun to receive attention in fabricating functional materials such as piezoelectric polymers, ceramics, and composites [34] despite sharing the same underlying principles as SLA. DLP adopted a distinct approach to pattern exposure, curing entire layers in a single step rather than point-by-point. In terms of print speed, layer uniformity, process scalability, and the functionalization of piezoelectric devices, this is a main difference between SLA and DLP.

4.4.1 Working principle of DLP

The 3D printing using DLP, a digital projector used to solidify liquid photopolymers; hence, the lamp in an inkjet-based system is replaced. The projector records two-dimensional images from successive sliced layers under the control of a digital micromirror device, in which thousands or millions of tiny mirrors are quickly toggled between an on-state and an off-state to modulate light exposure on a pixel basis once the projected image is displayed on the surface of the resin, all areas that receive light cure, building an entire layer at once [34]. Like SLA, the build platform incrementally moves up after curing each layer, allowing uncured resin to flow in and re-cover the surface. DLP-based systems can generally be configured in a bottom-up manner (Figure 4.2), with curing occurring against a clear window in a vat.

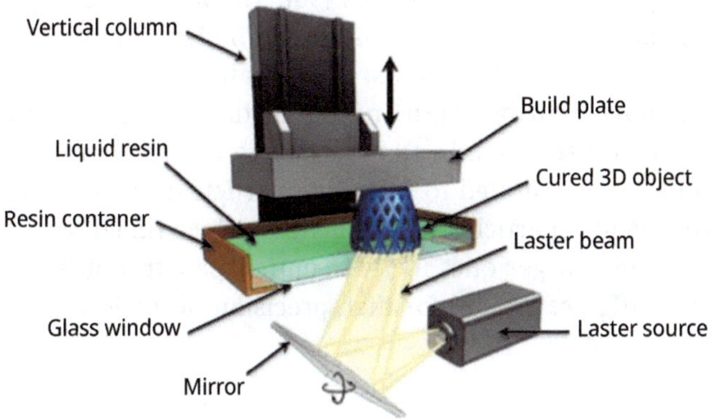

Figure 4.2: Schematic diagram of the bottom-up DLP process.

While this setup enables efficient resin use and higher printing resolution, peel forces and exposure settings must be carefully adjusted to ensure optimal results. The operation principle of DLP-based printing for piezoelectric materials, along with the layer-wise projection schematics and interplay with the resin vat and build platform, is depicted in Figure 4.2.

4.4.2 Piezoelectric materials for DLP

DLP-based piezomaterial systems utilize essentially the same materials as those in SLA, but their behavior with projected light has unique processing implications. Whole-layer exposure of resin is well-suited for high-loading, ceramic-filled resins, provided optical scattering losses are adequately addressed.

4.4.2.1 Piezoelectric polymers

PVDF-based copolymers, such that they can undergo photocuring in resins, which are their composites, are DLP printable. By repeated access to DLP, a consistent curing intensity can be achieved within each layer, resulting in reduced dimensional variability and less variation in polymer cross-linking between local areas. Such properties make them attractive for the fabrication of thin films, membranes, and microstructured piezoelectric devices for sensors and flexible electronics [35].

4.4.2.2 Ceramic-filled resins

DLP has been particularly attractive for printing ceramic piezoelectric materials, such as PZT, $BaTiO_3$, PVDF, and lead-free alternatives [36]. The layer-level light-exposure method enables rapid fabrication of a dense green body with homogeneous solidification, even at high ceramic contents. Light scattering by ceramic particles can also lead to a shallower curing depth and less distinct edges if the exposure conditions are not correctly selected. In DLP, factors to be considered for resin formulation include, but are not limited to, viscosity, particle dispersion stability, and optical clarity. The smaller particles yield a better surface finish, yet introduce more light scattering, while larger ones increase optical penetration at the cost of resolution. These trade-offs are crucial for high-precision piezoelectric ceramic components.

4.4.2.3 Composite and architectured materials

DLP enables the creation of architecturally designed piezoelectric composites with controlled porosity, lattice structures, and graded material distributions. These architectures may further enhance the piezoelectric response by reducing stiffness and improving active strain transfer [7]. These architecturally designed materials are being increasingly investigated for energy harvesting, acoustic sensing, and bioengineering applications.

4.4.3 Process parameters and exposure handling

The DLP-based printing is limited by exposure time, light intensity, pixel resolution, and layer thickness as the primary process parameters. The exposure time influences the polymerization rate and interlayer adhesion, while the intensity determines the curing depth and feature resolution. Pixel resolution, as set by the projector and optics system, specifies the minimum feature size that can be formed in the *XY* plane. Whereas, for SLA, the energy delivered from point to point depends locally on scan speed and hatch spacing, in DLP, every part receives uniform exposure over each given layer [4]. It simplifies process control through the uniformity of all its pixels, while exhibiting some pixelation on curved surfaces in space. The influence of the curing depth follows a similar model in DLP, as in SLA, even though the spatial energy distribution differs. Table 4.3 presents the DLP process parameters and their effects on the structural and functional properties of piezoelectric materials [3].

4.4.4 Post-processing and functional activation

Post-processing of DLP-printed piezoelectric materials is similar to that of SLA-printed piezoelectric materials. Three-dimensional printed parts are cleaned of uncured resin with a solvent, then UV post-cured to achieve complete polymerization [10]. For ceramic parts, binder removal and sintering are necessary to obtain dense, functional ceramics, which can provide a relatively homogeneous microstructure in green components, resulting in higher sinter density and lower defect levels. Electric poling is performed after sintering to activate piezoelectricity. The piezoelectric constants obtained are size, pore volume, and domain orientation dependent.

Table 4.3: DLP key parameters and their effect on the printed piezoelectric materials.

S. no.	DLP parameter	Typical optimized value/ range	Effect on printed parts
1	Layer thickness	10–20 (25 is common)	Thinner layers improve *Z*-direction resolution, reduce surface roughness, and improve printing accuracy. Thicker layers (>100 μm) can reduce interlayer adhesion, leading to lower density and piezoelectric constants.
2	Exposure time	2–50 s per layer (20 s is often optimal for dense PZT)	Insufficient time results in undercuring, leading to interlayer delamination and low green density. Excess time causes over-curing, reducing precision (light scattering) and inducing shrinkage stress.
3	Light intensity	5–6 mW/cm^2 (405 nm UV)	High intensities (e.g., >20 mW/cm^2) are beneficial for 0–3 piezoelectric composites, improving curing depth and resolution.
4	Slurry solid loading	45–75 wt% (or >50 vol% for ceramic)	Higher ceramic loading (up to ≈75 wt% or 55 vol%) increases the final sintered density and d_{33} value, and also significantly increases viscosity, affecting spreadability.
5	Print orientation	0° (parallel to base) or vertical (90°)	Parallel (0°) printing typically offers the best tensile strength; however, 90° (vertical) printing can improve compression strength and interlayer bonding by reducing "pancake" type defects.
6	Base layer and cure	3–5 layer; 30–40 s base exposure	Ensures strong adhesion to the build platform, preventing premature detachment, without significantly affecting final functional properties.
7	Wait time/ lift speed	5–10 s (wait); low speed (retract)	Longer wait times allow the high-viscosity ceramic resin to relax more fully, reducing voids and ensuring uniform layer thickness.

4.4.5 Advantages and disadvantages of DLP for piezoelectric materials

The most attractive features of DLP are its fast printing speed and layer uniformity. Because it cures entire layers at once, DLP also significantly reduces build times on large cross-sectional parts, in particular. Uniform exposure also promotes higher repeatability and dimensional consistency, an advantage for large-scale piezoelectric device manufacturing [1, 34]. DLP systems are typically limited in size by the native projector's resolution, or they trade off build volume for resolution. The artifacts introduced by pixelation can degrade the smoothness and quality of the geometry, particularly in curved or tilted features. Despite these limitations, DLP has been demonstrated to be an effective system in cases where higher speed, control, and uniformity are more critical than architectural considerations.

4.5 Comparison between SLA and DLP for piezoelectric materials

SLA and DLP are based on vat photopolymerization, but differences in exposure techniques, process control, and upscaling result in different outcomes when used with piezoelectric materials. Knowing the basics of these two processes allows a choice of manufacturing route to be determined based on the material system, device architecture, and function requirements. In this section, we compare SLA and DLP based on criteria critical to piezoelectric material fabrication, resolution, printing speed, microstructural homogeneity, material compatibility, and electromechanical performance. Table 4.4 presents a brief comparison between SLA and DLP for piezoelectric material fabrication, listing their pros and cons [37].

Table 4.4: Comparison between SLA and DLP for piezoelectric materials.

S. no.	Feature	SLA	DLP
1	Light source	UV laser (cure layer by layer)	Digital projector (flashes entire layer)
2	Build speed	Slow to medium (due to layer sintering)	Faster (cures the whole layer at once)
3	Accuracy	Excellent, very smooth surfaces	Moderate, rough surface
4	Build size	Small to large	Small to medium
5	Hardware cost	Medium to high	Medium to high
6	Materials	Broad range of high-performance resins	Broad range of high-performance resins
7	Application	Master pattern, functional prototype, jigs	Production parts, small-detailed parts, fit, functional parts

4.5.1 Resolution and geometric fidelity

Resolution is an essential feature of piezoelectric materials because geometric precision closely relates to the strain distribution and electric field uniformity, and accordingly, device efficiency. For SLA, the *XY*-plane resolution is determined by the laser spot size in projection and by the scanning precision [38]. Vector scanning in SLA ensures smooth contouring and high-resolution reproduction of curved and slanted features, particularly useful for complex piezoelectric geometries, such as those found in graded-thickness membranes and freeform actuators. In DLP, resolution is determined solely by the pixel size of the projected image. Although DLP provides uniform resolution across the entire build plane, the distinct pixel exposure can cause step effects or pixelation on rounded shapes. SLA is generally better than DLP for components with a predominantly planar or lattice-like architecture; however, DLP resolution is usually sufficient and readily reproducible.

4.5.2 Printing speed and productivity

The difference in printing speed between SLA and DLP is one of the most significant differences. In SLA, each layer is created by sweeping a laser across the entire cross section, so build time scales with geometric complexity and part size. Therefore, the SLA method is not suitable for scaling piezoelectric ceramic components in large or high-filling-ratio layers. DLP solves this limitation by fully curing each layer in a single exposure, rendering the building time almost insensitive to geometric complexity, which is particularly advantageous for ceramic-loaded piezoelectric resins, where a high solid loading increases the exposure requirement [39]. When it comes to producing piezoelectric components of the same design in large quantities or quickly prototyping multiple design iterations, there is no doubt that DLP has a clear productivity edge.

4.5.3 Compatibility of materials and processing

A wide variety of piezoelectric materials, including polymers, ceramics, and composites, have been investigated in both SLA and DLP. Their compatibility with heavily filled ceramic resins is markedly different. For SLA, laser energy can be locally adjusted by varying scan speed and power to provide fine control over curing depth in optically challenging resins [40]. This flexibility renders SLA less sensitive to resins that scatter light to a great extent. DLP is faster, but exposure time and power need to be optimized for each layer to achieve complete curing. The curing depth and edge sharpness can be substantially reduced when high ceramic loadings are used, thereby limiting the processing [41].

4.5.4 Microstructural homogeneity and functional performance

Microstructural homogeneity is a key factor in the functional properties of piezoelectric materials. Differences in density, porosity, and grain structure can lead to a non-uniform piezoelectric response, resulting in unreliable results. DLP layer-wide exposure leads to more uniform curing and a more uniform green density, resulting in microstructures that are more consistent across specimens after sintering [42]. In SLA and DLP, there are minor differences in exposure energy across a layer, especially for complex geometries. Although these variations are generally minor, they can lead to differences in shrinkage during sintering. Yet, the better surface finish and feature definition of SLS might improve electric-field homogeneity and reduce stress concentrations in piezoelectric devices. The d_{33} is the most essential property and depends on material composition, density, and domain orientation. Some studies have claimed that the d_{33} values of SLA and DLP-processed piezoelectric ceramics are, within measurement uncertainty, symmetric with those of their conventionally prepared counterparts after suitable post-processing [43].

4.5.5 Scalability and application suitability

Scalability is another important consideration. Note that SLA systems can also be scaled to build volume , with the ability to increase laser travel without any resolution loss, thus allowing the manufacturing of significant or nonuniform piezoelectric components [44]. DLP systems are limited instead by projector resolution, and larger build areas generally require reducing XY resolution or implementing more complicated optical system architectures. SLA is suitable for high-precision, low-volume piezoelectric devices, including microactuators, biomedical sensors, and research prototypes. For such applications, DLP is more ideal due to its fast fabrication and high repeatability, such as for dental piezoelectric elements, energy harvesters, and batch-fabricated sensor arrays.

4.5.6 Technology selection guidelines

The decision on whether to use SLA or DLP for piezoelectric manufacturing must consider factors such as resolution, speed, material properties, and specific application requirements. The SLA works well when surface finish, smooth curvature, and geometric freedom are required. DLP is beneficial for high-throughput, mass production with moderate build volumes, and uniform material properties. In various research and industrial areas, these two technologies are complementary rather than competitive [45].

4.6 The application of SLA and DLP-printed piezoelectric materials

SLA and DLP have expanded the application field of piezoelectric materials by enabling the creation of complex, high-resolution structures. In contrast to traditional processing routes, where geometry and integration are limited, vat photopolymerization enables personalized architectures, internal porosity, and multifunctional designs that immediately affect piezoelectric performance.

4.6.1 Application of SLA and DLP in sensors and actuators

Piezoelectric sensors and actuators are among the most developed applications of AM piezoelectric materials [17]. For sensors, mechanical stimuli such as pressure, vibration, or strain are transduced into electrical signals; in actuators, this process is reversed. SLA and DLP enable the fabrication of miniaturized sensors with complex structures, such as more curved surfaces and embedded channels, which are otherwise challenging to achieve with existing methods. For actuator use, geometrical freedom enables flexural structures, bending beams, and multilayered constructs that enhance the output of displacement and force. SLA high surface finish is beneficial for making thin-walled actuators, for which electric-field uniformity is a crucial issue. With its fast-layer curing, DLP enables high-volume sensor arrays and repetitive actuator elements.

4.6.2 Application of SLA and DLP in energy harvesting

Piezoelectric energy harvesting has been extensively studied for converting ambient mechanical energy into electrical power and is widely used in self-powered electronic devices and wireless sensor networks [16]. SLA and DLP enable the fabrication of architectured piezoelectric components, such as lattices with cores, porous beams, and graded-density elements, to enhance strain concentration and energy conversion. Porous and lattice-type designs reduce effective stiffness and improve mechanical compliance, enabling greater deformation at low frequencies. DLP is well-suited for the fabrication of such architecturally structured structures because of its layer-by-layer, consistent curing, and high throughput. These functionalities enable the construction of small, application-oriented energy-harvesting devices.

4.6.3 Application of SLA and DLP in biomedical and healthcare applications

Piezoelectric materials are commonly used in biomedical devices, including ultrasonic transducers, implantable sensors, and therapeutic applications [5, 46]. SLA- and DLP-based AM processes offer significant advantages for patient-specific customization and miniaturization. Complex shapes and patient-specific geometries can be manufactured directly from medical image data, reflecting the anatomical structure. SLA and

DLP-printed polymer piezoelectric materials are especially desirable for wearable and implantable devices due to their lightweight nature and low stiffness. Ceramic mechanical elements are stiffer but are used in high-performance ultrasound and diagnostic systems.

4.6.4 Application of SLA and DLP in microelectromechanical systems and microfluidics

The high resolution of SLA and DLP is well-suited for piezoelectric microelectromechanical systems (MEMS) and microfluidics. The use of piezoelectric elements in microfluidic chips enables control of fluid flow, mixing, and droplet generation. The capability to print enclosed channels and integrated functional components in a single stage of the production streamlines device manufacturing. SLA is generally well suited for smooth internal surfaces of channels. DLP can rapidly produce standard microfluidic components without molding, making it the chosen printing method. These approaches enable the fast prototyping and functional testing of piezoelectric MEMS.

4.6.5 Application of SLA and DLP in structural health monitoring and intelligent structures

Piezoelectric materials are frequently used in structural health monitoring systems to detect cracks, delamination, and vibration. SLA and DLP fabrication enable the fabrication of custom piezoelectric sensors that conform to complex structural surfaces, which are conductive not only to enhance signal sensitivity but also to track direct evidence of such activity across extensive spatial extents. Similarly, if piezoelectric transducers are integrated into printed structures, the processing of innovative materials and the integration of self-sensing components could also be realized. The fabricated composite piezoelectric designs, fabricated via vat photopolymerization, are capable of serving both load-bearing and sensing purposes, aligning with novel directions in intelligent materials and structures.

4.6.6 Application of SLA and DLP in electronics, acoustics, and beyond

In addition to traditional fields, piezomaterials based on SLA and DLP are in high demand for acoustic devices, flexible electronics, and human-machine interfaces. These are custom-shaped acoustic resonators, vibration-controlling components, and haptic feedback devices. For the above-mentioned product categories, vat photopolymerization allows a high degree of design freedom. New areas of application include soft robotics with piezoelectric actuators, which offer precise motion control, and wearable electronics that require lightweight, flexible sensors. The possibility of incorporating the piezoelectric effect directly into complex geometries enables SLA and DLP as enabling technologies for next-generation devices.

4.7 Emerging progress and research trends on SLA- and DLP-based piezoelectric fabrication

In the last few years, significant progress has been made in piezoelectric materials using SLA and DLP, driven by rapid advancements in materials science, photopolymer chemistry, and process engineering. From a technology limited to molecular-level polymer structures, 3D printing now offers the capabilities of an AM platform for structuring high-performance piezoelectric devices in an architecturally and multifunctional fashion.

4.7.1 High-speed continuous vat photopolymerization

One of the main research directions is to increase printing speed while maintaining resolution and functional properties. In typical vat photopolymerization processes, such as SLA and DLP, which involve discrete layer-by-layer separation from their curing station, continuously vat photopolymerization processes maintain an oxygen-inhibition zone near their vat window. This method of production enables continuous construction of the part, dramatically reducing construction time [41]. In piezoelectric materials, continuous printing helps minimize interlayer interfaces that can behave as mechanical and electrical discontinuities. Preliminary investigations indicate that the 24/7 DLP-based printing method yields increased uniformity of green bodies and reduced defect content in ceramic-substituted systems, resulting in enhanced piezoelectric performance following sintering and poling [43].

4.7.2 Sophisticated resin compositions for piezoelectric devices

Material composites constitute a significant domain of research. Innovative photocurable resins incorporating functional monomers, surface-modified ceramic particles, and specifically engineered photoinitiator systems have been formulated to enhance curing performance and compatibility. Surface modification of the piezoelectric ceramic particles improves dispersion stability and reduces light scattering, thereby further broadening the printable solid loading range. The development of lead-free piezoceramic materials, such as KNN and bismuth-based ceramics, is receiving increased research attention due to environmental concerns, health hazards, and regulations regarding the use of lead-based systems. SLA and DLP provide high-definition composition and architecture, perfect for the model of enhancement in lead-free piezoelectric performance.

4.7.3 Architectured and metastructured piezoelectric materials

Material architectures are among the most promising research directions enabled by vat photopolymerization. By programming periodic lattices, graded porosity, and topology-optimized structures, mechanical compliance and strain localization can be tuned to enhance piezoelectric sensitivity. These metastructured piezoelectric materials often exhibit coupled behavior, resulting in higher harvesting efficiency than their conventional dense bulk counterparts. DLP is especially suitable for printing these structures due to its homogeneous illumination and fast print speed. SLA is still applicable for parts with smooth curvature and fine feature control.

4.7.4 Multi-material and functionally graded printing

In multi-material printing, multiple materials are deposited onto the build platform during each print process, ensuring all materials are utilized. Another development is multi-material printing, which combines piezoelectric material with conductive, dielectric, or structural materials in a single component [47]. However, a significant challenge in vat photopolymerization is resin cross-contamination, which has recently been addressed by sequential resin exchange and localized curing, yielding promising results. Functionally graded piezoelectric material (FGPM), in which the composition and porosity are spatially varying, offers better impedance matching ratio properties, less stress concentration effect, and biased electromechanical response compared to the uniform FGPM. SLA and DLP provide the spatial resolution required to achieve such gradients with high accuracy.

4.7.5 Integration with simulation and data-driven design

The computational modeling of SLA and DLP fabrication techniques is becoming increasingly relevant to the design of piezoelectric materials. The distribution of the electric field, the stress response, and the energy conversion efficiency in complex printed geometries are all theoretically estimated using finite-element modeling. These models are used to support architectural design and inform parameter choice. Data-driven methodologies, such as machine learning, are being used to optimize resin recipes and exposure parameters, as well as to improve post-curing [48]. By correlating process parameters with piezoelectric properties, researchers aim to expedite material development and minimize the number of experimental iterations.

4.7.6 Reliability, aging, and long-term performance

With the prospect of applying AM piezoelectric materials, one current hot research focus is long-term reliability. Aging, fatigue resistance, and environmental stability studies are crucial for applications in structural monitoring, biomedical devices, and energy harvesting. Vat photopolymerization-derived piezoelectric materials should demonstrate reliable electromechanical properties over cyclic loading, temperature changes, and wet conditions. Mechanisms of degradation, as well as possible paths to mitigate damage, form part of active research.

4.8 Bibliometric analysis for the keyword "piezoelectric material fabricated by 3D printing"

The bibliometric analysis was performed using the Scopus database over the last 20 years for the keyword "piezoelectric material fabricated by 3D printing." The Scopus database reported 139 documents (05 January 2026). The .csv file was exported and opened in VOSviewer to construct a bibliometric network at an occurrence level of 6; 4,700 terms were identified, of which 118 met the threshold. For each of 118 terms, a relevance score was generated. Based on this score, the 60% most relevant terms were selected, i.e., 71 terms, as shown in Table 4.5. The network diagram is generated based on Table 4.5 and shown in Figure 4.3.

Table 4.5: Relevance and occurrence score for the keywords "piezoelectric material fabricated by 3D printing."

Id	Term	Occurrences	Relevance score
1	3D printing technology	11	1.1586
2	Ability	13	1.1017
3	Actuator	12	1.3966
4	Addition	13	0.7971
5	Advantage	15	0.9681
6	Array	7	0.8129
7	Barium titanate	14	1.5718
8	Batioinf3	9	1.1752
9	Batioinf3 inf	13	1.235
10	Bone regeneration	6	2.1333
11	BTO	9	1.3313
12	Ceramic	20	1.0428
13	Composite	28	0.3362
14	Defect	11	0.629
15	Density	21	0.7237
16	Diameter	10	1.2486
17	Dielectric property	6	0.9304
18	Energy harvester	14	0.4166
19	Extrusion	12	0.5776
20	Fabrication process	8	0.4704
21	Factor	15	1.2086
22	FDM	8	0.9212
23	Feasibility	8	0.7985
24	FFF	6	1.3299
25	Filament fabrication	6	1.3299
26	Force	13	0.8714
27	Frequency	17	1.2456
28	Geometry	20	0.7288
29	Influence	15	0.7263
30	Lead zirconate titanate	10	0.6402
31	Load	10	0.3597
32	Low cost	11	1.8661
33	Manufacturing	36	0.1734
34	Material property	12	0.5164
35	Mechanical property	10	0.8798
36	MEMS	8	1.7743
37	Multi-material	6	1.4631
38	Need	12	1.3637
39	Order	7	0.3577
40	Output voltage	9	0.4332
41	Pcn	22	1.0324
42	PDMS	8	0.8741
43	Peak	8	0.8377
44	Piezoelectric ceramic	9	1.1837
45	Piezoelectric coefficient	16	1.783

Table 4.5 (continued)

Id	Term	Occurrences	Relevance score
46	Piezoelectric property	22	0.4634
47	Piezoelectric response	9	0.3248
48	Piezoelectric sensor	14	0.4896
49	Piezoelectricity	11	0.8979
50	PLA	8	0.7089
51	Poly	19	1.0262
52	Polydimethylsiloxane	8	0.8741
53	Polyvinylidene fluoride	9	1.0634
54	Possibility	8	1.1708
55	Pressure	10	0.8354
56	Printer	12	1.5365
57	PVDF	15	1.1872
58	Range	23	1.1916
59	Resolution	9	2.3131
60	Resonant frequency	8	2.4214
61	Sample	18	0.641
62	Self	10	0.949
63	SEM	8	0.6408
64	Stereolithography	11	0.9969
65	System	35	1.072
66	Temperature	12	0.9719
67	Term	7	0.9055
68	Transducer	16	0.6818
69	Use	17	0.5869
70	Vinylidene fluoride	9	1.7486
71	Way	8	0.5456

A research gap analysis for the keyword "3D printing technology" was conducted, and related research is highlighted in Figure 4.4. The 3D printing processes used for the fabrication of piezoelectric materials were SLA and fused deposition modeling (FDM).

A research gap analysis for the keyword "stereolithography" was conducted, and related research is highlighted in Figure 4.5.

A section of a keyword cloud showing reported research on piezoelectric property for 3D printing as shown in Figure 4.6.

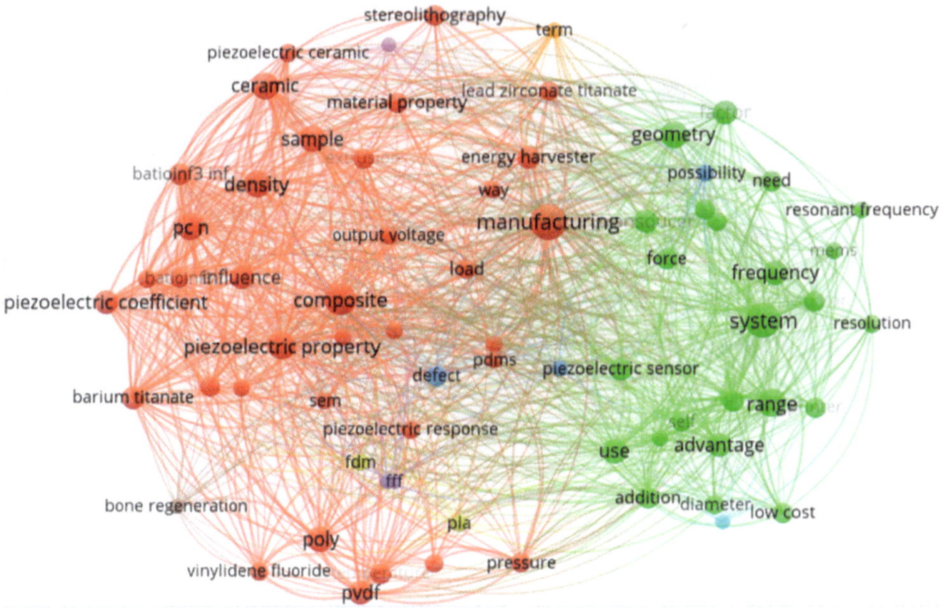

Figure 4.3: The bibliometric network diagram for the keyword "piezoelectric material fabricated by 3D printing."

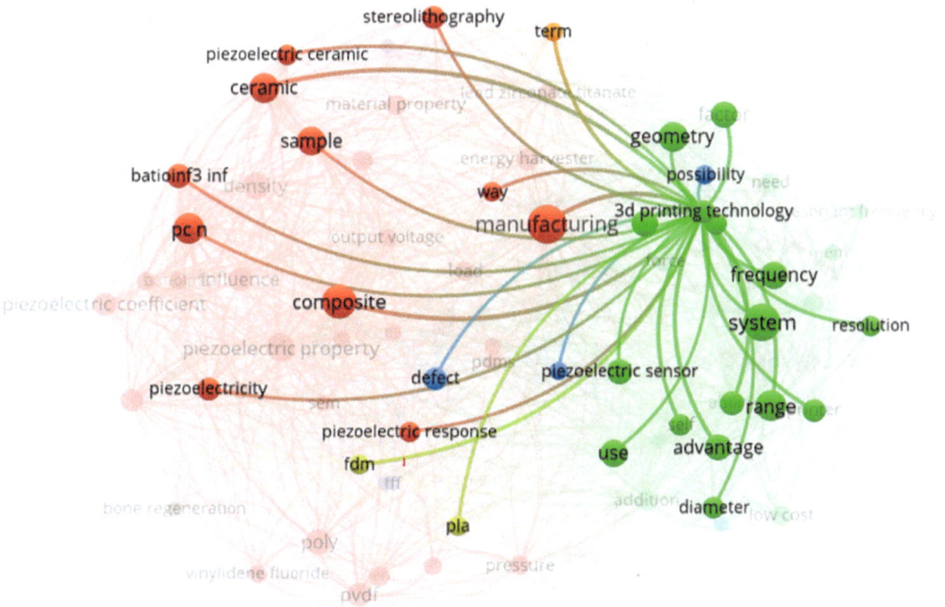

Figure 4.4: The research gap analysis for the node of "3D printing technology."

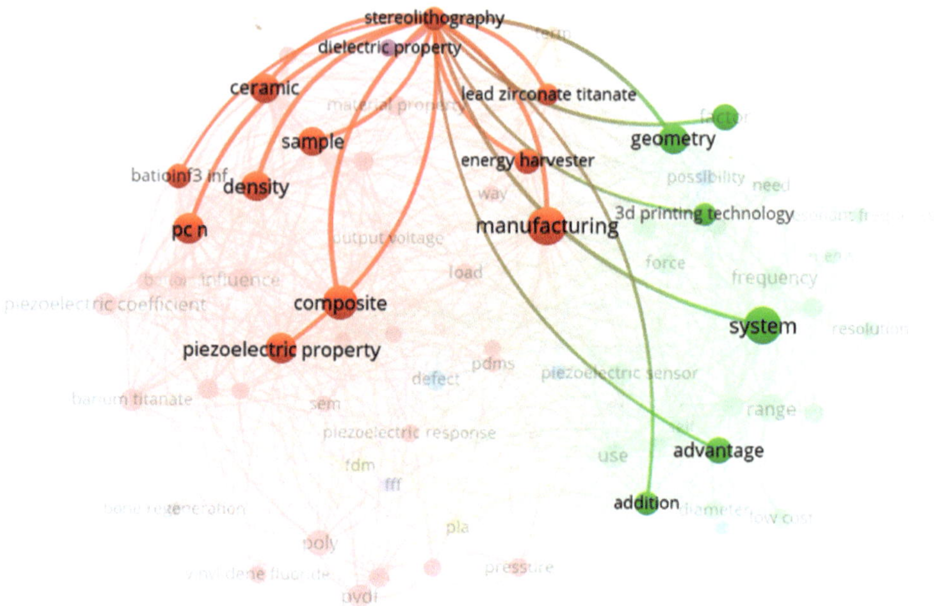

Figure 4.5: The research gap analysis for the node of "stereolithography."

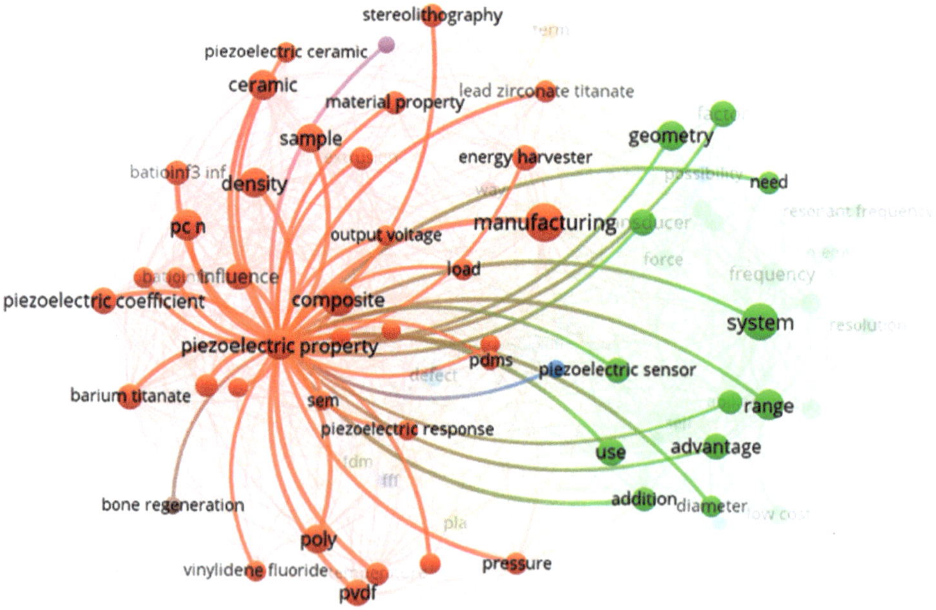

Figure 4.6: The research gap analysis for the node of "piezoelectric property."

4.9 Conclusions

- SLA and DLP are powerful AM technologies for piezoelectric material-based structural applications, enabling overall structural designs and functionalities that were previously impossible.
- By appreciating the structure, composition, and microarchitecture of 3D photo-patterned poled polymers, these methods overcome several difficulties associated with traditional piezoelectric fabrication methods.
- The basic principles of SLA and DLP, including material formulation approaches, process parameters, and post-processing techniques suitable for piezoelectric applications, were outlined throughout this chapter.
- The comparison studies show that, although the SLA offers superior surface finishing and greater flexibility in exposure control, the DLP achieves a faster build rate and better layer curing; they are therefore complementary systems rather than competitors.
- The application spectrum of AM piezoelectric materials is also growing, including sensors, actuators, energy harvesters, biomedical devices, MEMS, and innovative structures.
- Latest developments in the field of architected, multi-material piezoelectric devices produced by vat photopolymerization indicate that such performance can be fully met, and even exceeded, compared to conventional fabrication.
- SLA- and DLP-based 3D printing technologies have the potential to be transformative tools for advancing piezoelectric materials and devices in the next generation. Their further improvement can not only expand the design space of piezoelectric technology but also facilitate their applications across emerging disciplines, from engineering to healthcare and energy.

References

[1] Wang H, Lai Q, Zhang D, Li X, Hu J, Yuan H. A Systematic Study on Digital Light Processing 3D Printing of 0-3 Ceramic Composites for Piezoelectric Metastructures. Research. 2025a Feb 21;8:0595.

[2] Han D, Yang C, Fang NX, Lee H. Rapid multi-material 3D printing with projection micro-stereolithography using dynamic fluidic control. Additive Manufacturing. 2019 May 1;27:606–15.

[3] Cahyono SI, Arifvianto B, Suyitno SUA, Prihandana GS, Yusof F, Sing SL, Narayan RL, Mahardika M. Fabrication of Metallic Products Using Digital Light Processing-Based Additive Manufacturing: A Review. Advanced Engineering Materials. 2026 Jan;28(1):e202402845.

[4] Mohanya S, Krishnanand NA, Taufik M. Estimation of surface roughness for digital light processing-based additively manufactured parts. Progress in Additive Manufacturing. 2025 May;10(5):3779–98.

[5] Wang X, Shujaat S, Shaheen E, Jacobs R. Quality and haptic feedback of three-dimensionally printed models for simulating dental implant surgery. The Journal of Prosthetic Dentistry. 2024 Apr 1;131 (4):660–67.

[6] Yunus DE, Sohrabi S, He R, Shi W, Liu Y. Acoustic patterning for 3D embedded electrically conductive wire in stereolithography. Journal of Micromechanics and Microengineering. 2017 Mar 14;27 (4):045016.
[7] Schoonraad SA, Fischenich KM, Eckstein KN, Crespo-Cuevas V, Savard LM, Muralidharan A, Tomaschke AA, Uzcategui AC, Randolph MA, McLeod RR, Ferguson VL. Biomimetic and mechanically supportive 3D printed scaffolds for cartilage and osteochondral tissue engineering using photo-polymers and digital light processing. Biofabrication. 2021 Sep 16;13(4):044106.
[8] Shujaat S, Da costa senior O, Shaheen E, Politis C, Jacobs R. Visual and haptic perceptibility of 3D printed skeletal models in orthognathic surgery. Journal of Dentistry. 2021 Jun 1;109:103660.
[9] Shin J, Tabatabaei Rezaei N, Choi S, Li Z, Kim DH, Kim K. Photocrosslinkable Kidney Decellularized Extracellular Matrix-Based Bioink for 3D Bioprinting. Advanced Healthcare Materials. 2025 Sep;14 (24):2501616.
[10] Szalai S, Szívós BF, Nemes V, Szabó G, Kurhan D, Sysyn M, Fischer S. Investigation of Digital Light Processing-Based 3D Printing for Optimised Tooling in Automotive and Electronics Sheet Metal Forming. Journal of Manufacturing and Materials Processing. 2025 Jan 15;9(1):25.
[11] Kolibaba TJ, Killgore JP, Caplins BW, Higgins CI, Arp U, Miller CC, Poster DL, Zong Y, Broce S, Wang T, Talačka V. Results of an interlaboratory study on the working curve in vat photopolymerization. Additive manufacturing. 2024 Mar 25;84:104082.
[12] Yu PY, Su YC. UV Polymerized 3D Semipermeable Structures for Rapid Osmotic Pumping. In 2022, the 21st International Conference on Micro and Nanotechnology for Power Generation and Energy Conversion Applications (PowerMEMS) 2022 Dec 12 (pp. 204–07). IEEE.
[13] Husain M, Singh R, Pabla BS. Effect of meta-structure on mechanical properties of polyvinylidene fluoride composite-based 3D printed intramedullary pins. Journal of Thermoplastic Composite Materials. 2024a Oct;37(10):3319–39.
[14] Dou R, Tang W, Hu K, Wang L. Ceramic paste for space stereolithography 3D printing technology in microgravity environment. Journal of the European Ceramic Society. 2022 Aug 1;42(9):3968–75.
[15] Kumar R, Kumar V, Husain M, Singh R, Kumar P. On Fabrication of Novel Polyvinylidene Fluoride BaTiO3-Cu-Doped ZnO-Based Filament for 3D Printing of Pacemakers. Journal of Materials Engineering and Performance. 2025 Jul 7;1–4.
[16] Yang C, Song S, Chen F, Chen N. Fabrication of PVDF/BaTiO3/CNT piezoelectric energy harvesters with bionic balsa wood structures through 3D printing and supercritical carbon dioxide foaming. ACS Applied Materials & Interfaces. 2021 Aug 25;13(35):41723–34.
[17] Mansour R, Omoniyi OA, Reid A, Liang L, O'Leary R, Windmill JF. A Novel 3D-Printed (0-3) Piezocomposite Material for Sensing Applications. In 2020, IEEE International Conference on Flexible and Printable Sensors and Systems (FLEPS) 2020 Aug 16 (pp. 1–4). IEEE.
[18] Kim K, Zhu W, Qu X, Aaronson C, McCall WR, Chen S, Sirbuly DJ. 3D optical printing of piezoelectric nanoparticle–polymer composite materials. ACS nano. 2014 Oct 28;8(10):9799–806.
[19] Ribeiro C, Costa CM, Correia DM, Nunes-Pereira J, Oliveira J, Martins P, Gonçalves R, Cardoso VF, Lanceros-Méndez S. Electroactive poly (vinylidene fluoride)-based structures for advanced applications. Nature protocols. 2018 Apr;13(4):681–704.
[20] Wang W, Sun J, Guo B, Chen X, Ananth KP, Bai J. Fabrication of piezoelectric nano-ceramics via stereolithography of low viscous and non-aqueous suspensions. Journal of the European Ceramic Society. 2020 Mar 1;40(3):682–88.
[21] Chen Z, Song X, Lei L, Chen X, Fei C, Chiu CT, Qian X, Ma T, Yang Y, Shung K, Chen Y. 3D printing of piezoelectric element for energy focusing and ultrasonic sensing. Nano Energy. 2016 Sep 1;27:78–86.
[22] Husain M, Singh R, Pabla BS. On dynamic reduced order model for the flexural characteristics of polyvinylidene fluoride composite-based functional prototypes. Journal of Materials Engineering and Performance. 2025a Jun;34(11):9714–23.

[23] Husain M, Singh R, Pabla BS. On Prominent β-Phase of PVDF Matrix with 4D Capabilities. Journal of The Institution of Engineers (India). 2025b Feb;106(1):341–52.
[24] Yuan X, Yan A, Lai Z, Liu Z, Yu Z, Li Z, Cao Y, Dong S. A poling-free PVDF nanocomposite via mechanically directional stress field for self-powered pressure sensor application. Nano Energy. 2022 Jul 1;98:107340.
[25] Xie M, Zhu M, Yang Z, Okada S, Kawamura S. Flexible self-powered multifunctional sensor for stiffness-tunable soft robotic gripper by multimaterial 3D printing. Nano Energy. 2021 Jan 1;79:105438.
[26] Gaytan SM, Cadena MA, Karim H, Delfin D, Lin Y, Espalin D, MacDonald E, Wicker RB. Fabrication of barium titanate by binder jetting additive manufacturing technology. Ceramics International. 2015 Jun 1;41(5):6610–19.
[27] Sitthi-Amorn P, Ramos JE, Wangy Y, Kwan J, Lan J, Wang W, Matusik W. MultiFab: a machine vision-assisted platform for multi-material 3D printing. ACM Transactions on Graphics (Tog). 2015 Jul 27;34 (4):1–1.
[28] Cheng J, Chen Y, Wu JW, Ji XR, Wu SH. 3D printing of BaTiO3 piezoelectric ceramics for a focused ultrasonic array. Sensors. 2019 Sep 20;19(19):4078.
[29] Wang ZH, Zhang YZ, Miao WJ, Wu FB, Wang SQ, Ouyang JH, Wang YM, Zou YC. Vat Photopolymerization-Based Additive Manufacturing of Si3N4 Ceramic Structures: Printing Optimization, Debinding/Sintering, and Applications. Materials. 2025b Mar 29;18(7):1556.
[30] Song S, Li Y, Wang Q, Zhang C. Boosting piezoelectric performance with a new selective laser sintering 3D printable PVDF/graphene nanocomposite. Composites Part A: Applied Science and Manufacturing. 2021 Aug 1;147:106452.
[31] Liu CL, Du Q, Zhou H, Chen S, Wu JM, Zhang G, Shi YS. 3D printing of lead zirconate titanate piezoelectric ceramics via digital light processing (DLP). Ceramics International. 2023 Sep 1;49 (17):28492–99.
[32] De Marzo G, Fachechi L, Antonaci V, Mastronardi VM, Portaluri L, Todaro MT, Algieri L, Qualtieri A, Rizzi F, Scaraggi M, De Vittorio M. On the measurement of piezoelectric d33 coefficient of soft thin films under weak mechanical loads: A rapid and affordable method. Materials & Design. 2024 Nov 1;247:113399.
[33] Tao R, Shi J, Granier F, Moeini M, Akbarzadeh A, Therriault D. Multi-material fused filament fabrication of flexible 3D piezoelectric nanocomposite lattices for pressure sensing and energy harvesting applications. Applied Materials Today. 2022 Dec 1;29:101596.
[34] Tiller B, Reid A, Zhu B, Guerreiro J, Domingo-Roca R, Jackson JC, Windmill JF. Piezoelectric microphone via a digital light processing 3D printing process. Materials & Design. 2019 Mar 5;165:107593.
[35] Du Y, Wang R, Zeng M, Xu S, Saeidi-Javash M, Wu W, Zhang Y. Hybrid printing of wearable piezoelectric sensors. Nano Energy. 2021 Dec 1;90:106522.
[36] Florczak S, Lorson T, Zheng T, Mrlik M, Hutmacher DW, Higgins MJ, Luxenhofer R, Dalton PD. Melt electrowriting of electroactive poly (vinylidene difluoride) fibers. Polymer International. 2019 Apr;68 (4):735–45.
[37] Francesco M, Di Fiore A, Gobbato E, Fioretti A, Zuccon A, Stellini E. Comparison between SLA and DLP printing materials in dentistry: a review. In Atti 27 °Congresso nazionale Collegio dei Docenti Universitari di discipline Odontostomatologiche 2020 (Vol. 12, pp. 403–04).
[38] Kim H, Torres F, Islam MT, Islam MD, Chavez LA, Rosales CA, Wilburn BR, Stewart CM, Noveron JC, Tseng TL, Lin Y. Increased piezoelectric response in functional nanocomposites through multiwall carbon nanotube interface and fused-deposition modeling three-dimensional printing. MRS communications. 2017 Dec;7(4):960–66.

[39] Suaste-Gómez E, Rodríguez-Roldán G, Reyes-Cruz H, Terán-Jiménez O. Developing an ear prosthesis fabricated in polyvinylidene fluoride by a 3D printer with sensory intrinsic properties of pressure and temperature. Sensors. 2016 Mar 4;16(3):332.
[40] Nassar H, Khandelwal G, Chirila R, Karagiorgis X, Ginesi RE, Dahiya AS, Dahiya R. Fully 3D printed piezoelectric pressure sensor for dynamic tactile sensing. Additive Manufacturing. 2023 Jun 5;71:103601.
[41] Maines EM, Porwal MK, Ellison CJ, Reineke TM. Sustainable advances in SLA/DLP 3D printing materials and processes. Green Chemistry. 2021;23(18):6863–97.
[42] Xue R, Yuan P, Zhao B, Jing F, Kou X, Yue W, Wang D, Sewvandi GA, Hu D. DLP printing of BT/HA nanocomposite ceramic scaffolds using low refractive index BT crystals. Journal of Materiomics. 2024 Sep 1;10(5):1036–48.
[43] Coppola B, Lacondemine T, Tardivat C, Montanaro L, Palmero P. Designing alumina-zirconia composites by DLP-based stereolithography: Microstructural tailoring and mechanical performances. Ceramics International. 2021 May 15;47(10):13457–68.
[44] Borrello J, Nasser P, Iatridis JC, Costa KD. 3D printing a mechanically-tunable acrylate resin on a commercial DLP-SLA printer. Additive manufacturing. 2018 Oct 1;23:374–80.
[45] Truxova V, Safka JI, Seidl M, Kovalenko I, Volesky LU, Ackermann MI. Ceramic 3D printing: Comparison of SLA and DLP technologies. MM Science Journal. 2020 Jun 10; 2020:3905–11.
[46] Singh R, Husain M, Mehra S. On 3D printing of functionally graded implants using two different metallic powder-based alloys. National Academy of Sciences Letters. 2024 Aug;47(4):369–74.
[47] Hussain MI, Xia M, Ren X, Ge C, Jamil M, Gupta MK. Digital light processing 3D printing of ceramic materials: a review on basic concept, challenges, and applications. The International Journal of Advanced Manufacturing Technology. 2024b Jan;130(5):2241–67.
[48] Hassanpour M, Narongdej P, Alterman N, Moghtadernejad S, Barjasteh E. Effects of post-processing parameters on 3D-printed dental appliances: A review. Polymers. 2024 Oct 1;16(19):2795.

Minhaz Husain*, Ranvijay Kumar, Rupinder Singh

Chapter 5
Inkjet and direct ink writing of a functional piezoelectric structure

Abstract: Although extensive research has been conducted on piezoelectric materials over the past two decades, the development of functional piezoelectric structures using additive manufacturing techniques, particularly inkjet printing and direct ink writing (DIW) remains relatively limited. In this chapter, an investigation of inkjet printing and DIW for the fabrication of functional piezoelectric structures has been conducted from a process-capability perspective. Recently, and over the past decades, multimaterial additive manufacturing (AM) technologies have been developed, offering new possibilities by allowing engineering materials with selected properties to be continuously mixed and tailored to application requirements, enabling the manufacture of complex geometries and multifunctional components. Inkjet printing and DIW are the basis for commercial printing techniques due to their accuracy, versatility, and adaptability to a wide range of feedstock materials. This technique utilizes an extrusion-based process to deposit viscoelastic inks with specific rheological properties. The process is compatible with many material classes, including polymers, ceramics, and composite systems, making this approach highly attractive for functional and biomedical applications. This chapter also includes a case study of the assessed process capability of both techniques, using the dielectric constant as the key performance index. Inkjet-printed polyvinylidene fluoride-50% graphite composites maintain stable dielectric properties with minimal deviation, demonstrating high process reliability. Results suggest that DIW-fabricated thermoplastics, such as polycaprolactone (PCL), and room-temperature-vulcanized (RTV) liquid silicone rubber (LSR) displayed consistent dielectric characteristics within their respective limitations, thereby validating the fidelity and reproducibility of the process.

Keywords: 3D printing, inkjet printing, direct ink writing, functional structures, piezoelectric materials

---

***Corresponding author: Minhaz Husain**, Marwadi University Research Center, Mechanical Engineering Department, Marwadi University, Rajkot, Gujarat, India, e-mail: minhazme2012@gmail.com
Ranvijay Kumar, Marwadi University Research Center, Mechanical Engineering Department, Marwadi University, Rajkot, Gujarat, India
Rupinder Singh, Mechanical Engineering Department, National Institute of Technical Teachers Training and Research, Chandigarh, India

 | https://doi.org/10.1515/9783112218846-005

5.1 Introduction

In the modern era, utility engineering materials have emerged to meet a range of engineering and nonengineering requirements. These engineering materials are well designed for processing using recent technologies, such as 3D printing. 3D printing enables the manufacturing of functional structures using these engineering materials [1, 2]. The piezoelectric materials offer tailored performance and fulfill application needs when processed via 3D printing. In recent years, inkjet and direct ink writing (DIW)-based 3D printing processes have emerged with the potential to manufacture piezoelectric material-based functional structures [3–5].

Inkjet 3D printing has emerged as one of the established methods for the manufacturing of functional structures using binders, polymers, and resins [6, 7]. Some 3D printing processes that use the inkjet printing principle are material jetting, binder jetting, and nanoparticle jetting. Inkjet printing is also known for its high accuracy in the manufacturing of parts, with excellent surface finish, color printing, high speed, and the ability to print complex parts [8, 9]. However, there are some limitations to inkjet printing that must be addressed. The materials used in this process are typically expensive, which in turn makes production costly. Also, compromised mechanical strength, increased postprocessing requirements, and higher printer costs are among the drawbacks of this technique. For piezoelectric materials, injection printing is one of the most important techniques, with several studies reported in recent years [10, 11]. Inkjet printing enables the manufacturing of devices such as energy harvesters, energy generators, sensors, and actuators [11]. Inkjet printing can manufacture fully printed devices for electronic applications [12]. Due to excellent processing capabilities and the ability to fabricate complex parts, previous studies have reported the manufacture of microelectromechanical system-based piezoelectric transducers for medical and biological applications [13]. Similarly, previous studies have reported the inkjet printing of piezoelectric materials for a variety of applications, including thin films for transparent haptics [14], sensors for self-powered monitoring [15], and actuators [16].

DIW is also an additive manufacturing technique that belongs to the material extrusion processes. In the DIW process, the layers are typically solidified through drying, photo-curing, cooling, and chemical reactions. A wide range of materials is used in the DIW process, including ceramics, polymers, hydrogels, conductive inks, particle suspensions, food formulations, and bioinks loaded with living cells [4, 17, 18]. The use of these materials in the DIW process depends on several factors, including shear-thinning properties, shape-retention behavior, and solidification rate. DIW is one of the acceptable manufacturing methods due to several advantages, such as the ability to use a variety of materials, support multimaterial printing, use low-cost machinery, and manufacture parts at low cost [18, 19]. For these reasons, the DIW process is widely used in food printing, energy storage and harvesting, electronics, and biomedical engineering [20–23]. In recent years, several studies have also reported the fabrication of functional structures using piezoelectric materials. It was reported that DIW of aqueous-based lead zirconate titanate lanthanum (PLZT) was demonstrated for potential applications such as underwater acoustic transducers [24]. Li et al. [22] reported the DIW of geometrically complex parts for sensing and actuation applications. Similarly, several studies have been reported in recent years on the development of functional structures using piezoelectric materials [20, 24–27].

5.2 Inkjet printing for the manufacturing of functional structures of piezoelectric materials

The inkjet printing applies principles of traditional inkjet printing to an industrially conventional 3D fabrication process. It does this by dispensing small droplets of matter, or binding agents, laboriously, layer by layer, into complex geometries directly from digital computer-aided design (CAD) files [28], which can take place through two fundamental processes: material jetting and binder jetting. Both processes utilize controlled droplet formation and allow precise spatial localization, followed by solidification or bonding after the droplets fall onto the substrate, making them suitable for a wide range of engineering and biomedical applications. First, design the CAD model in SolidWorks, and then process it in the slicer software [29]. That model is then cut into thin layers, which, when stacked, form a cross-section of the final object. These slices direct the movement of the printhead and lay out where each layer is deposited. Most printheads are composed of arrays of picoliter-range micro-nozzles that eject droplets via thermal or piezoelectric actuation mechanisms. Such mechanisms control droplet size, speed, and localization during deposition with extreme precision, enabling the fabrication of high-resolution-based structures [19].

Material jetting is similar to inkjet printing, in which droplets of liquid photopolymers are directly deposited onto the build platform. Immediately after deposition, these droplets are exposed to ultraviolet (UV) light, which initiates a photopolymerization reaction that rapidly solidifies the liquid portion [30]. This deposition and simultaneous curing enable the fabrication of complex structures up to a few meters in dimension with high-fidelity detail in shape, surface finish, and resolvable fine features. Once a layer is completed, the build platform will be lowered by a predetermined distance, and the next layer will be deposited or cured; this process continues until the object is completed. This technique is particularly advantageous for applications that require precision and the capability to print multiple materials simultaneously [31].

Binder jetting is a process in which a liquid binding agent is applied to the powder bed of desired metals, ceramics, or polymers in selective areas. On the build platform, a uniform layer of powder is deposited, allowing the print head nozzles to dispense binder droplets at specified locations defined by the desired geometry. The droplets penetrate the powder, forming localized particle bonding. After each pass, powder is deposited, and the process repeats layer by layer. As a consequence, shaped architectural components are formed and referred to as green parts, which usually require postprocessing, such as drying, to develop the desired mechanical properties. This technology is especially favored for its broad material compatibility and its ability to process complex geometries quickly.

Inkjet-based 3D printing is governed by multiple crucial factors, such as droplet formation dynamics, wetting, and solidification interactions, which are at play during its working mechanism. Therefore, droplet generation must be controlled, as it can cause defects such as satellite droplets and nonuniform deposition. In binder jetting, the interaction between the binder and the powder bed is extremely significant, as it affects the strength and integrity of each printed layer. The researchers then classified the defect mechanisms observed in binder jetting as sources of poor final part quality, partitioning them into powder spreading, layer formation, and binder distribution [30]. While this mechanism enhances mechanical strength and durability, it also has limitations, such as shrinkage and distortion. 3D inkjet printing is the most advanced technology, offering significant versatility across various solution spaces. For example, in the pharmaceutical manufacturing context, binder jetting has been explored to print patient-specific drug formulations with defined dosage and release profiles, advancing precision medicine. The properties of powder materials and process parameters have enhanced mechanical properties and sustainability across multiple applications [31].

The inkjet 3D printing technology, which achieves high resolution and surface quality through in situ curing of ejected materials, is illustrated in Figure 5.1. Due to powder systems and postprocessing techniques, binder jetting enables the use of a broader range of materials. This method, however, faces several challenges when printing with inkjet, including defect generation and material interactions during process optimization. Continued research will make these technologies more effective and reliable, and they may become staples of advanced manufacturing.

5.3 DIW for the manufacturing of functional structures of piezoelectric materials

DIW is an extrusion-based additive manufacturing (AM) technique that builds 3D structures by precisely depositing viscoelastic inks. In DIW, an ink or gel is continuously extruded through a nozzle and deposited layer by layer to form complex geometries. DIW is governed by the interactions among ink rheology, extrusion dynamics, and postdeposition stabilization [32]. This specific ability makes DIW one of the most versatile methods in AM, enabling fabrication from polymers, ceramics, hydrogels,

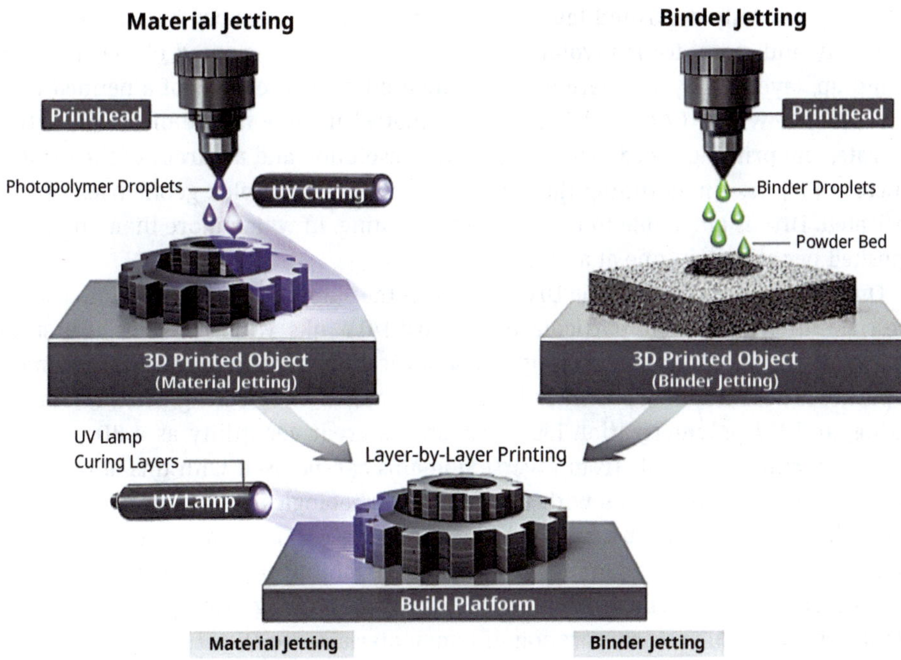

Figure 5.1: Schematic of the inkjet 3D printing process.

and composite systems, which begins by preparing a printable ink with finely tuned rheological properties. Shear-thinning is not only a basic requirement but also results in a decrease in viscosity with increasing applied shear stress during extrusion, followed by rapid recovery when the stress is released [33], giving the material a great degree of freedom to flow through the nozzle and be molded into a filament on the bed. The ink had to have a finite yield stress so it could support its own weight. Among these, one key is the relationship between rheological behavior and ink printability; a poor choice of ink properties can lead to deformation, spread, or collapse of the printed structure [34].

Once the ink is ready, it is transferred to a reservoir, which can be a syringe-based system or an attached laser-thin nozzle. Extrusion is pneumatic or mechanical, in which controlled pressure forces the ink from the nozzle. Constrained fluid mechanics governs flow within the nozzle, with viscosity, shear rate, and nozzle diameter influencing extrusion behavior [35]. Among them are uniform and constant flow. DIW faces challenges, such as nozzle clogging and flow instability, that are strongly dependent on the interaction between ink composition and flow conditions [36]. When the ink leaves the nozzle, a continuous filament is created and deposited onto a substrate or previous layers. The ink stability is governed by its viscoelasticity, surface tension, and the trace of its movement on the underlying layer. If the ink is too runny, it may bleed too much and reduce resolution; if it is too solid, it may fail to cohere with the previously deposited layer [37]. Thus, having the proper balance between flowability and structure is pivotal for successful prints. 3D-printed object from the ground up, layer by layer, where each extruded ink forms sections of a defined path in accordance with a CAD model. These parameters include nozzle diameter, extrusion rate, and printing speed, which impact the resolution and accuracy of the printed structure [38]. By finely tuning these parameters, more complex geometries can be fabricated. DIW is amenable to multimaterial printing, in which more than one ink is deposited per structure, one at a time.

One important feature of the DIW process is the stabilization of the printed structure right after deposition, particularly for most DIW inks, which are soft at first (or semi-solid) and often require subsequent solidification steps to improve mechanical properties. These mechanisms include solvent removal, heat curing, chemical cross-linking, and UV polymerization [39]. DIW shows great versatility as a 3D printing method, as many materials from diverse domains can be used with this approach if they are processed into inks with the required rheological behavior. It has been shown that materials can be printed from polymers, ceramics, and more advanced nanocomposites, as well as conductive inks. This capability enables a variety of applications across the biomedical engineering, electronics, energy storage, and environmental systems sectors. Incorporation of nanomaterials into DIW inks is one method that has been shown to significantly enhance the structural properties, electrical conductivity, and functional performance of the resulting 3D-printed structures [4, 40].

While DIW is executing on its promises, it continues to face numerous challenges. This heavy reliance on ink rheology results in elaborate material formulations, often tailored to a specific application. To ensure uniform extrusion and minimize defects such as ink fracture or misalignment, the process parameters need to be precisely controlled. In addition, linking material properties to printing conditions makes it difficult to establish standardized protocols. It involves the coordination of ink formulation, extrusion, ink deposition during printing, and stabilization after deposition, which are directly comparable to those on other dialogue platforms. Despite challenges in optimizing material behavior and process parameters, DIW is still developing as a powerful, flexible AM with extensive reach across multiple sectors of science/industry.

5.4 Case studies for inkjet printing and DIW of functional structures

As reported by Mehta et al. 2024 [41], the gel printing of polyvinylidene fluoride (PVDF)-graphite-based functional structures [41]. The study highlighted research on the sensing, dielectric, and mechanical properties of PVDF-50% graphite composites. The findings of this study confirm that the aforementioned material combination is compatible with gel printing and relevant for piezoelectric sensing applications. The PVDF-50% graphite sample showed a dielectric constant of 20.49. The dielectric constant must be measured with precision during processing, as it plays a key role in sensor design. The process capability analysis was conducted by repeating the experiments (on recommended settings) 15 times in this study. The composition of the PVDF-50% graphite composite dielectric values is shown in Table 5.1.

Table 5.1: Dielectric observation of the PVDF-50% graphite.

Exp.	Dielectric constant of PVDF-50% graphite
1	20.53
2	20.45
3	20.47
4	20.56
5	20.45
6	20.65
7	20.45
8	20.59
9	20.44
10	20.35
11	20.51
12	20.49
13	20.55
14	20.56
15	20.39

An illustration of the process capability analysis for structuring PVDF-50% graphite based on its dielectric constant is presented in Figure 5.2, which roughly resembles a normal distribution curve with specification limits and the target value overlaid. The resultant data is tightly clustered around the target dielectric constant of 20.49, indicating that this process is well calibrated and yields consistent results. A narrow distribution of data around the mean reflects high repeatability and low process variance. It is worth noting that all the values measured schematically are within the lower and upper specification limits, thus confirming that the process meets all quality requirements. One of the key assumptions in process capability analysis is that the distribution is approximately normal, or that variations are random rather than systematic. Leakage of this type is a sign of bad process capability. In addition to being accurate, the process is precise, as the data points are tightly clustered. Such precision is critical for applications with dielectric materials, as even slight deviations can dramatically affect performance.

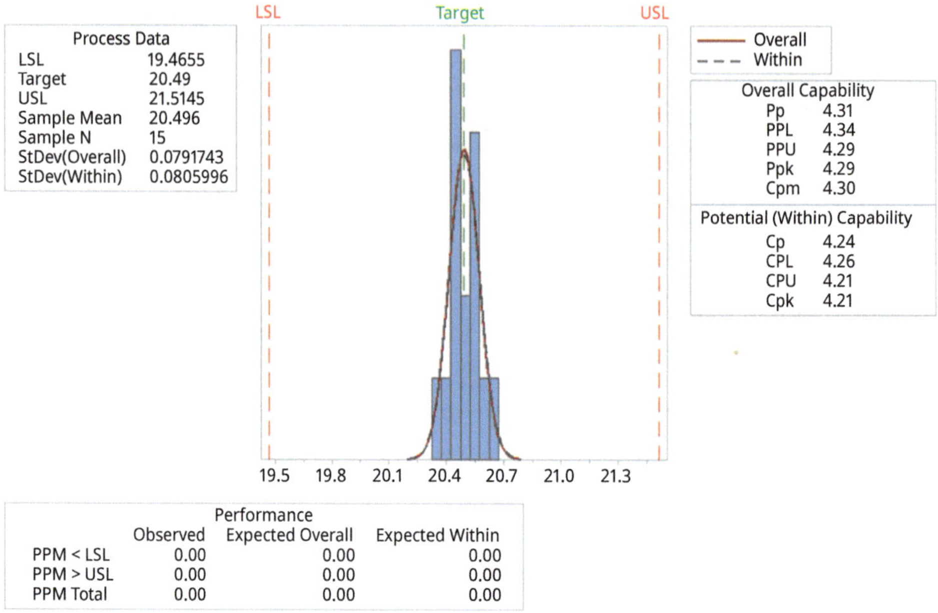

Figure 5.2: Process capability for the dielectric constant of PVDF-50% graphite.

DIW is an established process for developing the functional structure of piezoelectric materials. This process enables the fabrication of sensors and sensor parts with effective process control. To investigate the process capability parameters of a DIW-based functional structure, a process capability analysis was performed on polycaprolactone (PCL) and room-temperature-vulcanized (RTV)-liquid silicone rubber (LSR)-based parts. The suggested dielectric constants were 3.65 and 9.81 for PCL- and RTV-SLR-printed parts, respectively. The 15 repeated experiments were again performed to investigate the process capability analysis.

The dielectric constant values of the PCL-based functional structure are given in Table 5.2. Dielectric constant of PCL structures fabricated by DIW, as determined from Table 5.2. The experimental results range from 3.59 to 3.71 across 15 experiments, which is close to the designed value of 3.65. The upper and lower specification limits are 3.4675 and 3.8325, respectively, and all measured values lie within the acceptable range, indicating that the DIW process exhibits excellent stability and repeatability during PCL printing. The relatively small variation in dielectric constant indicates that the flow and deposition characteristics of the material generated during each run of the extrusion-based printing process remain fairly consistent. Low values are naturally population averages, and slight variations in nozzle pressure, ink rheology, or specific printing conditions can lead to deviations from the target. But these variations are negligible and do not affect the material's performance. The dielectric constant values around the target indicated a well-centered process, critical to achieving consistent dielectric properties for functional components. As a thermoplastic polymer, PCL shows stable dielectric properties, which are beneficial for applications in flexible electronics and biomedical devices. The data shown in this table demonstrate that DIW can yield PCL-based constructs with reliable, repeatable dielectric properties.

Table 5.2: Dielectric observation of the PCL.

Exp.	Dielectric constant of PCL
1	3.67
2	3.68
3	3.65
4	3.68
5	3.60
6	3.59
7	3.71
8	3.65
9	3.66
10	3.59
11	3.61
12	3.70
13	3.63
14	3.69
15	3.64

Process capability [35, 36] for the dielectric constant of PCL-based structures produced using DIW (Figure 5.3), which is likely a distribution of measurement values within the given limits and target. The data seem to be compacted around 3.65, suggesting the process can be centered and deliver a close result. The relatively low data spread indicates high precision and minimal variability across the printing process. All values are within the lower and upper specification limits, indicating the process is in control. It will follow a normal distribution, suggesting that the variation is random (not systematic), which is a shortcoming of the methodology.

If a high degree of overlap between the data distribution and the target value is observed, it indicates good process capability with little risk of creating out-of-spec components, particularly critical for applications that require consistent dielectric properties across multiple parts. The moving average was necessary to demonstrate that DIW processes, including extrusion rate, nozzle movement, and material consistency, are controlled and governed. Figure 5.3 shows that the DIW process for PCL is reproducible and efficient; therefore, it could be used for manufacturing parts that require stable dielectric performance.

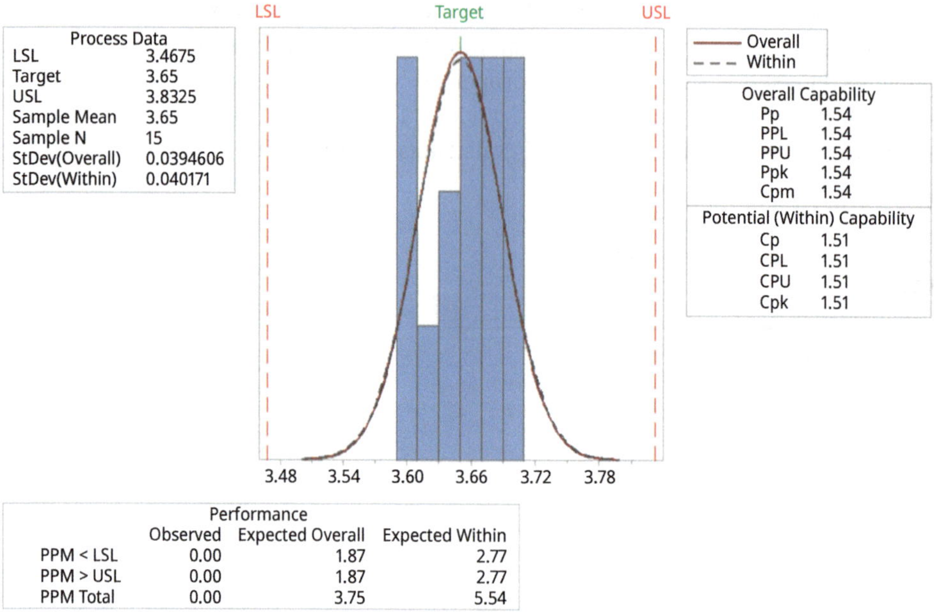

Figure 5.3: Process capability for the dielectric constant of PCL.

Table 5.3 depicts the dielectric constant of RTV-LSR structures prepared using DIW. The corresponding variable values, i.e., 9.72–9.91, closely match the target value of 9.81. Note that the lower and upper specification limits are 9.3195 and 10.3005, respectively, and that all experimental values fall within this range, indicating a well-controlled process that leads to reproducible dielectric properties. The small spread of measured values indicates a stable process, though the dispersion is slightly wider than PCL due to the viscoelasticity of silicone-based materials. Dielectric response can be affected by the progression of curing, material flow, and environmental conditions. The results show that these factors are well controlled in the process, and the process is well centered. RTV-LSR is well-suited for flexible electronics and soft sensors due to its elastic and dielectric properties. The results in Table 5.3 confirm that DIW produces RTV-LSR objects with well-replicable dielectric properties. Overall, the process is robust and appropriate for producing silicone-based functional materials.

Figure 5.4 depicts the process capability analysis of the dielectric constant for DIW-fabricated RTV-LSR structures, which is most likely a statistical picture of the measured values relative to specification limits/target values. The data appear clustered around the target dielectric constant of 9.81, suggesting the process is well calibrated. Silicone-based inks and their curing behavior are included in the distribution; it is slightly broader than for other materials due to inherent variability.

All dielectric constant values are within specifications, indicating that the process can produce a product that meets quality requirements. Noticeable variations are

Table 5.3: Dielectric observation of the RTV-LSR.

Exp.	Dielectric constant of RTV-LSR
1	9.82
2	9.91
3	9.90
4	9.89
5	9.81
6	9.80
7	9.77
8	9.85
9	9.78
10	9.72
11	9.89
12	9.77
13	9.88
14	9.76
15	9.75

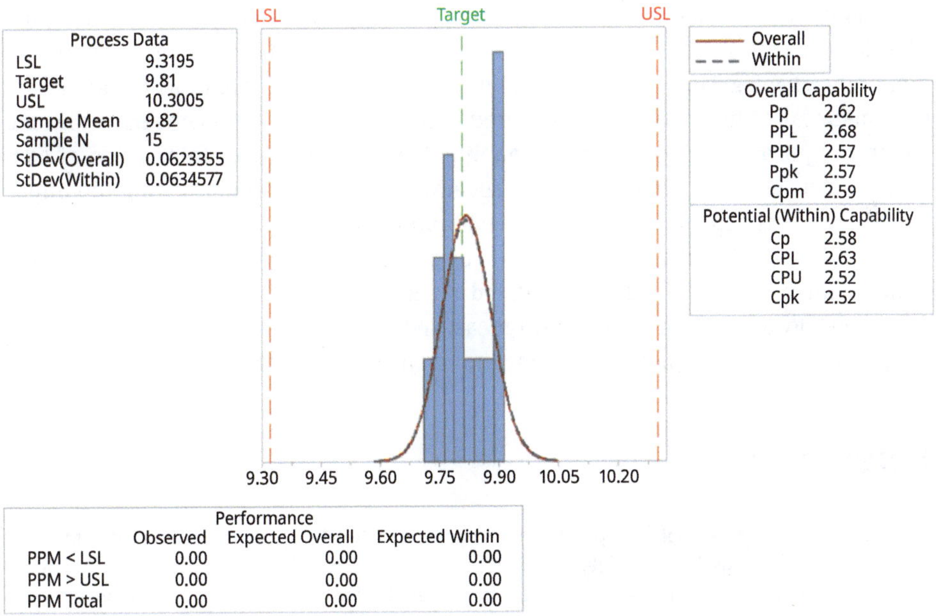

Figure 5.4: Process capability for the dielectric constant of RTV-LSR.

simply random and manageable. Good process centering is indicated by the data being close to the target value for the first variable, while enough spread indicates acceptable variability. As shown, while RTV-LSR may introduce additional challenges due to material properties, DIW remains capable of providing reliable, consistent outcomes, especially relevant for applications that require flexible, durable dielectric materials. Figure 5.4 shows that the combination of these process control strategies is rather effective in DIW. At the same time, RTV-LSR can be an adequate material for advanced functional applications.

5.5 Conclusions

The chapter concludes that inkjet printing and DIW are powerful AM techniques for fabricating functional piezoelectric structures. Inkjet and DIW processes offer benefits in precision, material compatibility, and design flexibility, making them ideally suited for developing complex parts that meet the desired specifications. Inkjet printing is an AM technology that can produce high-definition parts with a great surface finish and precise droplet placement (by material jetting or binder drawing). Basically, DIW is a highly versatile method for 3D printing thermoplastics, elastomers, and composites. Both extrusion and ink stability, layering, and adhesion depend on the optimization of the rheological properties of the ink used for writing. Due to its advantages in multimaterial printing, DIW is considered highly promising for engineering and biomedical applications, as well as for enabling low-cost, large-area fabrication of complex geometries. This chapter presents several case studies demonstrating the importance of process capability analysis for evaluating the robustness and reproducibility of AM processes. The dielectric constants of the structures were measured with both inkjet-printed and DIW. All investigated structures exhibited repeatability below unity. The close concordance between measured values and target specifications indicates good control over both processing and material behavior.

References

[1] Singh T, Kumar S, Sehgal S. 3D printing of engineering materials: A state of the art review. Materials today: proceedings. 2020 Jan 1;28:1927–31.
[2] Yan Q, Dong H, Su J, Han J, Song B, Wei Q, Shi Y. A review of 3D printing technology for medical applications. Engineering. 2018 Oct 1;4(5):729–42.
[3] De Gans BJ, Duineveld PC, Schubert US. Inkjet printing of polymers: state of the art and future developments. Advanced materials. 2004 Feb 3;16(3):203–13.
[4] Saadi MA, Maguire A, Pottackal NT, Thakur MS, Ikram MM, Hart AJ, Ajayan PM, Rahman MM. Direct ink writing: a 3D printing technology for diverse materials. Advanced Materials. 2022 Jul;34 (28):2108855.

[5] Singh M, Haverinen HM, Dhagat P, Jabbour GE. Inkjet printing—process and its applications. Advanced materials. 2010 Feb 9;22(6):673–85.

[6] Calvert P. Inkjet printing for materials and devices. Chemistry of Materials. 2001 Oct 15;13 (10):3299–305.

[7] Guo Y, Patanwala HS, Bognet B, Ma AW. Inkjet and inkjet-based 3D printing: connecting fluid properties and printing performance. Rapid Prototyping Journal. 2017 Apr 18;23(3):562–76.

[8] Mueller J, Shea K, Daraio C. Mechanical properties of parts fabricated with inkjet 3D printing through efficient experimental design. Materials & Design. 2015 Dec 5;86:902–12.

[9] Zub K, Hoeppener S, Schubert US. Inkjet printing and 3D printing strategies for biosensing, analytical, and diagnostic applications. Advanced Materials. 2022 Aug;34(31):2105015.

[10] Feng Y, Liu J, Deng J, Zhang S, Liu Y. A review on droplet-based 3D printing with piezoelectric micro-jet device. Smart Materials and Structures. 2024 Jul 1;33(7):073003.

[11] Thuau D, Kallitsis K, Dos Santos FD, Hadziioannou G. All inkjet-printed piezoelectric electronic devices: energy generators, sensors and actuators. Journal of Materials Chemistry C. 2017;5 (38):9963–66.

[12] Aliqué M, Simão CD, Murillo G, Moya A. Fully-printed piezoelectric devices for flexible electronics applications. Advanced Materials Technologies. 2021 Mar;6(3):2001020.

[13] Sumerel J, Lewis J, Doraiswamy A, Deravi LF, Sewell SL, Gerdon AE, Wright DW, Narayan RJ. Piezoelectric ink jet processing of materials for medical and biological applications. Biotechnology Journal: Healthcare Nutrition Technology. 2006 Sep;1(9):976–87.

[14] Glinsek S, Song L, Kovacova V, Mahjoub MA, Godard N, Girod S, Biagi JL, Quintana R, Schleeh T, Guedra M, Rupin M. Inkjet-printed piezoelectric thin films for transparent haptics. Advanced Materials Technologies. 2022 Aug;7(8):2200147.

[15] Abdolmaleki H, Haugen AB, Merhi Y, Nygaard JV, Agarwala S. Inkjet-printed flexible piezoelectric sensor for self-powered biomedical monitoring. Materials Today Electronics. 2023 Sep 1;5:100056.

[16] Brünahl J, Grishin AM. Piezoelectric shear mode drop-on-demand inkjet actuator. Sensors and Actuators A: Physical. 2002 Oct 30;101(3):371–82.

[17] Li L, Lin Q, Tang M, Duncan AJ, Ke C. Advanced polymer designs for direct-ink-write 3D printing. Chemistry–A European Journal. 2019a Aug 14;25(46):10768–81.

[18] Rocha VG, Saiz E, Tirichenko IS, García-Tuñón E. Direct ink writing advances in multi-material structures for a sustainable future. Journal of Materials Chemistry A. 2020;8(31):15646–57.

[19] Li M, Du W, Elwany A, Pei Z, Ma C. Binder jetting additive manufacturing of metals: a literature review. In International Manufacturing Science and Engineering Conference 2019b Jun 10 (Vol. 58745, p. V001T01A033). American Society of Mechanical Engineers.

[20] Hassan MS, Zaman S, Rodriguez A, Molina L, Dominguez CE, Morgan R, Bernardin J, Lin Y. Direct ink write 3D printing of wave propagation sensor. Flexible and printed electronics. 2022 Dec 1;7 (4):045011.

[21] Li Y, Li L, Li B. Direct ink writing of three-dimensional (K, Na) NbO3-based piezoelectric ceramics. Materials. 2015 Apr 14;8(4):1729–37.

[22] Li Z, Li J, Luo H, Yuan X, Wang X, Xiong H, Zhang D. Direct ink writing of 3D piezoelectric ceramics with complex unsupported structures. Journal of the European Ceramic Society. 2022 Aug 1;42 (9):3841–47.

[23] Liu L, Huang P, Xu S, Chen X, Fu K, Li J, Liu H, Xue W, Shao K, Wu S. Self-polarized cellulose nanofiber-reinforced PVDF-based piezoelectric composites via direct-ink-writing 3D printing for pressure sensing and energy harvesting. Sensors and Actuators A: Physical. 2025 Jan 1;381:116084.

[24] Liu K, Zhang Q, Zhou C, Shi Y, Sun C, Sun H, Yin C, Hu J, Zhou S, Zhang Y, Fu Y. 4D printing of lead zirconate titanate piezoelectric composites transducer based on direct ink writing. Frontiers in Materials. 2021 May 28;8:659441.

[25] Pinto RS, Serra JP, Barbosa JC, Gonçalves R, Silva MM, Lanceros-Méndez S, Costa CM. Direct-ink-writing of electroactive polymers for sensing and energy storage applications. Macromolecular Materials and Engineering. 2021 Nov;306(11):2100372.
[26] Shi X, Xu M, Zhang H, Sun Y, Wang R, Yuan Z, Liu H. Direct ink writing nanogenerator: Texture structure and its influence on the piezoelectric performance. Nano Research. 2025 Jun 1;18(6).
[27] White AM. Direct Ink Writing of Flexible Piezoelectric Sensors. Boise State University, 2022.
[28] Srivastava K, Kumar Y. Binder Jetting: A Review on Process Parameters and Challenges. Manufacturing and Processing of Advanced Materials. 2023 Dec;14:28–42.
[29] Zhao K, Su Z, Ye Z, Cao W, Pang J, Wang X, Wang Z, Xu X, Zhu J. Review of the types, formation mechanisms, effects, and elimination methods of binder jetting 3D-printing defects. Journal of Materials Research and Technology. 2023 Nov 1;27:5449–69.
[30] Wang Y, Genina N, Müllertz A, Rantanen J. Binder jetting 3D printing in fabricating pharmaceutical solid products for precision medicine. Basic & Clinical Pharmacology & Toxicology. 2024 Mar;134 (3):325–32.
[31] Miyanaji H, Rahman KM, Da M, Williams CB. Effect of fine powder particles on quality of binder jetting parts. Additive Manufacturing. 2020 Dec 1;36:101587.
[32] Wei P, Cipriani C, Hsieh CM, Kamani K, Rogers S, Pentzer E. Go with the flow: Rheological requirements for direct ink write printability. Journal of Applied Physics. 2023 Sep 14;134(10).
[33] Singh G, Singh R, Kumar C, Husain M. On characterization of room temperature vulcanizing-liquid silicone rubber with sensing capabilities for novel orthotic shoe sole. Materials Letters. 2025 Jun 15;389:138350.
[34] Husain M, Singh R, Singh M, Singh G. On fumed silica-reinforced liquid silicon rubber for a novel orthotic shoe sole. Materials Letters. 2025 Nov 15;139803.
[35] Singh G, Singh R, Husain M, Singh A. On-Process Capability Analysis of Novel Cerclage Strip for Health Monitoring of Comminuted Fracture in Canines. In Advanced Manufacturing Technologies in Biomedical Science: Practical Applications, Case Studies, and Future Trends 2026 (pp. 460–67). CRC Press.
[36] Husain M, Singh R, Pabla BS. On process capability of single screw extruder for fabricating PVDF composite matrix. Materials Today: Proceedings. 2023a Nov 10.
[37] Husain M, Singh R, Pabla BS. On PVDF composite prepared by single screw extruder: A statistical analysis of mechanical properties. Materials Today: Proceedings. 2023b Oct 29.
[38] Sauret A, Ray TR, Compton BG. Fluid mechanics challenges in direct-ink-writing additive manufacturing. Annual Review of Fluid Mechanics. 2026;58.
[39] Wilt JK, Gilmer D, Kim S, Compton BG, Saito T. Direct ink writing techniques for in situ gelation and solidification. MRS Communications. 2021 Apr;11(2):106–21.
[40] Tsang AC, Zhang J, Hui KN, Hui KS, Huang H. Recent development and applications of advanced materials via direct ink writing. Advanced Materials Technologies. 2022 Jul;7(7):2101358.
[41] Mehta A, Singh R, Pabla BS, Kumar R. On the recycling of PVDF-reinforced graphite by using 3D gel printing. Materials Letters. 2024 Jul 1;366:136536.

Gurwinder Singh, Rupinder Singh

Chapter 6 Selective laser sintering and powder bed fusion-based 3D printing for piezoelectric materials

Abstract: In the past few decades, numerous studies have reported on the use of selective laser sintering (SLS) and powder bed fusion (PBF) in biomedical and implant applications. However, little has been reported on SLS and PBF-based 3D printing for piezoelectric materials for biomedical and implant applications. This study presents the fabrication of compressive samples in such a way that the top and bottom layers are solid parts and the intermediate zone has three different meta-structures (octet-truss, honeycomb, and diamond) to evaluate the piezoelectric behavior of 17-4 precipitated-hardened stainless steel fabricated via PBF-laser beam (PBF-LB) for biomedical and implant applications. The study's findings suggest that three samples exhibit compressive strengths of 6,530, 5,155, and 6,610 N, indicating good mechanical strength for applications in fields where load-bearing capacity with bone is anticipated, such as orthopedics or biomedical applications. Also, the sample with octet-truss meta-structure showed a favorable outcome in terms of compressive strength near the natural bone of canine tibia, which further offers better performance and provides more accurate and reliable mechanical interaction with postoperative tibia behavior. Based on the volume fraction, phase analysis reveals that the majority of Body Centered Cubic (BCC) and Face Centered Cubic (FCC) phases contribute to microstructural stability, resulting in a strong, stable overall structure and higher sample strength. Furthermore, the results are supported by scanning electron microscopy, energy-dispersive spectroscopy, porosity %, and CALPHAD analysis.

Keywords: SLS, 3D printing, LPBF, biomedical, piezoelectric property, piezoelectric materials

6.1 Introduction

Selective laser sintering (SLS) has established itself as a highly versatile additive manufacturing (AM) process across numerous fields, including the pharmaceutical and biomedical engineering sectors, as well as the industrial sector [1]. Fina et al. [2]

Gurwinder Singh, Rupinder Singh, Mechanical Engineering Department, National Institute of Technical Teachers Training and Research, Chandigarh, India

 | https://doi.org/10.1515/9783112218846-006

demonstrated that porous, controlled-release tablets can be generated by tuning laser parameters, thereby indicating the applicability of SLS technology in the manufacturing of pharmaceutical dosage forms, enabling solvent-free and personalized medicine production. Awad et al. [3] provided a thorough analysis of SLS principles. They highlighted its advantages, including high resolution, greater design freedom for prospective features, and efficient manufacturing for customized drug delivery systems. Gueche et al. [4] stated that porosity % must be maintained in SLS-fabricated parts, as it obviously affects dissolution behavior and overall structural integrity, and is then used, for example, in precision-coordinated drug delivery. Furthermore, another study shows that SLS technology can be used to manufacture complex multi-drug systems using advanced formulation methods. But material scarcity remains an ongoing issue [5]. Another research outlined that SLS has been used in the bio- and engineering fields to produce personalized implants, prosthetics, and scaffolds for tissue regeneration, which has drawn significant attention to its importance for a patient-specific healthcare approach [6]. In the SLS, functional prototype-based implants or components are fabricated by sintering powdered material layer by layer, facilitating the fabrication of more complex geometries without support structures, which is beneficial in industrial, biomedical, aerospace, and tissue regeneration and cell growth applications with tunability in controlled porosity scaffolds [7]. Research by Rahmani et al. [8] indicates that SLS is a promising candidate for tissue regeneration and cell growth, as it enables the fabrication of biomaterials with controllable porosity during scaffold production. Another pharmaceutical study further demonstrates that SLS, which enables precise control over drug dosage and release kinetics, is an advanced method for designing a personalized medicine system [9]. Moreover, SLS has the great ability and flexibility to process a wide range of polymers and composites efficiently, making it more suited for the fabrication of functional prototyping for the automotive and aerospace sectors and thereby the adoption of SLS in engineering applications where digital fabrication and mass customization are favorable, such as in new fields of advanced manufacturing system, wearable sensors, and customized implants and consumer products [10, 11]. On the other hand, due to its low energy density and partial sintering process, SLS is also confined mainly to polymer-based materials, restricting its applications in load-bearing or high-strength structures. SLS is also less applicable to dense metallic elements because it primarily uses thermoplastic (nylon and polymer composites) materials [12]. To address this limitation, selective laser melting (SLM), a preeminent powder-bed fusion process for metals, has been introduced. Metal powders are fully melted, producing completely dense parts with superior mechanical properties [13]. A study by Kushwaha et al. [14] outlined that the evolution from SLS to SLM 3D printing plays a major role in developing precision metallic implants and load-bearing structures made from high-strength alloy-based materials, with tunable properties that require biocompatibility for biomedical, orthopedic, and veterinary applications. Gu [15] and Gao et al. [16] highlighted that SLM has become a highly developed PBF technique. Their processing parameters play a crucial role in shaping microstructure and mechanical properties, enabling the fabrication of dense metallic structures with complex topologies and tailored microstructures, and making it especially suitable for biomedical applications compared to conventional manufacturing processes. SLM is a commonly used technique in the fabrication of patient-specific implants and scaffolds due to its high precision, thereby improving the process and reducing costs by addressing concerns such as stress shielding and osseointegration by tailoring porosity and surface shape, enhancing implant integration with the bone tissue environment [17, 18]. Gatto et al. [19] outlined that optimal input parameters for stainless-steel (SS) alloys produced using DMLS play an important role in implant quality, including corrosion behavior and mechanical properties, which, in turn, impact the dependability, longevity, and life expectancy of the implant. Another study noted that the use of SLM-fabricated Ti and SS implants, as well as porous-based structures in several dental and orthopedic fields, provides improved biocompatibility, corrosion resistance, and strength while enhancing osseointegration and improving implant contact bonding with surrounding cells [20, 21]. Furthermore, Hassan et al. [22] developed biodegradable magnesium alloys produced by SLM. The researchers demonstrated the potential of these alloys for use as temporary implants, given their predictable degradation behavior and biocompatibility. Gracheva et al. [23] examined biodegradable metallic systems, such as magnesium, iron, and zinc. They emphasized that SLM can modify degradation kinetics and mechanical properties for next-generation implants, thereby extending beyond the scope of traditional materials. Furthermore, Li et al. [24] demonstrated that SLM can be used to tailor the superelasticity and shape-memory effects of NiTi shape-memory alloys, rendering these alloys candidates for smart biomedical devices. Ginestra et al. [25] described the general impact of SLM as compared to other AM methods. SLM offers improved surface properties, promotes better cell adhesion and osteointegration, and provides the longevity of the implants. Puebla et al. [26] demonstrated that direct metal laser sintering (DMLS) can produce patient-specific implants with improved bone ingrowth characteristics. This ability was highlighted, as DMLS enables the production of complex Ti-6Al-4V structures with precisely controlled porosity and very high mechanical strength. Traini et al. [27] reported improved osseointegration in DMLS-manufactured titanium dental implants with porous surface morphology, which encourages greater bone-implant contact than traditional implants. Van Noort [28] emphasizes the disruptive potential of DMLS in dentistry, enabling the fabrication of customized crowns, bridges, and implants with high precision, minimal waste generation, and short turnaround times. DMLS can produce dense, corrosion-resistant parts suitable for biomedical implants, and it has investigated the processing of biocompatible metals by laser-based powder-bed fusion [29]. Mazzoli [30] highlighted that one of the important applications of DMLS in the field of biomaterials, such as Ti-6Al-4V, is the manufacturing of scaffolds with controlled, interconnected porosity and customized flexibility that fulfill the needs of tissue engineering and regenerative medicine. Similarly, another study by Hrabe and Quinn [31] investigated its microstructure and mechanical characteristics and emphasized the importance of post-processing treatments for assured reliable implant performance. In a study on the influence of process parameters on DMLS titanium alloys, [32] reported that rapid solidification results in fine martensitic microstructures, which, in turn, lead to greater strength and hardness, making the material suitable for load-bearing applications in the biomedical sector. Sun et al. [33] reported that DMLS enables the production of porous titanium implants with greater flexibility and freedom, resulting in better biological response, enhanced cell adhesion and proliferation, and making them ideal for orthopedic applications. According to Bose et al. [34], DMLS can produce porous metallic implants that retain the characteristics of natural bone architecture while addressing challenges such as residual stresses and surface finishing, offering a broader perspective on biomaterials and their AM. Singh et al. [35] outlined the challenges of DMLS technology, including the manufacture of site-specific implants, limited material fabrication, waste of unused powder, lattice structures, control over surface roughness and porosity, the wrapping effect, and thermal distortion in the final part for biomedical applications.

The literature indicates that numerous studies have reported on the use of SLS and LPBF in biomedical and implant applications. However, little has been reported on SLS and PBF-based 3D printing for piezoelectric materials for biomedical and implant applications. This study presents a fabrication of compressive samples based on three different meta-structures (octet-truss, honeycomb, and diamond) to evaluate the piezoelectric behavior of 17-4 precipitated-hardened (17-4PH) SS fabricated via PBF-LB for biomedical and implant applications.

6.2 Materials and methods

For the bibliometric analysis, the SCOPUS database over the past 20 years was used to identify research gaps and trends using the keyword "selective laser sintering and powder bed fusion piezoelectric materials printing", yielding a total of 605 documents. Further, an open-source VOS viewer was used to build the bibliometric networks, with a minimum of 10 occurrences per term. Of the 6,536 terms, 179 met the threshold using the .csv file extracted from the SCOPUS database. Additionally, the 179 terms were selected based on their occurrence and relevance scores (Table 6.1) and used to highlight the network diagram shown in Figure 6.1. Figure 6.1 presents the network diagram based on the specific keyword "selective laser sintering and powder bed fusion piezoelectric materials printing." Figures 6.2 and 6.3 highlight the potential use of SLS and LPBF 3D printing across various fields, including biomedical implants, scaffolds, and tissue engineering applications.

Additionally, both Figures 6.2 and 6.3 highlight the research gap: limited studies have been conducted on selective laser sintering (SLS) and powder bed fusion (PBF)-based 3D printing for piezoelectric materials. This study highlights the fabrication of compressive samples (three different meta-structures: octet-truss, honeycomb, and diamond) as a pilot run using the LPBF process on 17-4PH SS material to evaluate piezo behavior, based on thermomechanical analysis, for biomedical applications. Figures 6.4 and 6.5 show the country and area-wise work reported on the utilization of SLS 3D printing across various fields, including biomedical and implant applications.

Table 6.1: Occurrence and relevance score based on a particular keyword.

ID	Term	Occurrences	Relevance score
1	3D printing	137	0.6555
2	3D printing technology	40	2.042
3	3DP	17	2.149
4	4D printing	54	1.2624
5	Ability	27	0.325
6	Addition	31	0.2914
7	Additive	14	4.0622
8	Additive manufacturing	247	0.082
9	Additive manufacturing process	15	0.3483
10	Additive manufacturing technology	19	0.364
11	Advance	25	0.1954
12	Advancement	46	0.1855
13	Advantage	50	0.1104
14	Aerospace	35	0.0829
15	Alloy	23	1.3655
16	AM process	25	0.4924
17	AM technique	18	0.4218
18	AM technology	13	0.6454
19	Analysis	68	0.2158
20	Application	264	0.2176
21	Area	47	0.4213
22	Art	19	0.4999
23	Article	42	0.2213
24	Aspect	24	0.326
25	Benefit	25	0.4402
26	Binder jetting	11	0.3247
27	Biocompatibility	19	0.9851
28	Biocomposite	14	1.6308
29	Biomaterial	23	1.3905
30	Book	22	1.9981
31	Ceramic	74	0.2894
32	Challenge	140	0.1863
33	Chapter	28	0.5784
34	Characteristic	51	0.2631
35	Combination	18	0.5515
36	Complex structure	14	0.1847
37	Component	85	0.6815
38	Composite	79	0.3429
39	Composite material	21	0.4815

Table 6.1 (continued)

ID	Term	Occurrences	Relevance score
40	Comprehensive overview	14	0.4671
41	Comprehensive review	28	0.2954
42	Conductive polymer composite	11	0.8108
43	Construction	18	0.4145
44	Cost	38	0.1776
45	Creation	19	1.5869
46	Current challenge	10	1.0466
47	Current state	13	0.371
48	Defect	36	0.5601
49	Demand	29	0.3269
50	Deposition	23	0.6027
51	Detail	17	0.5637
52	Development	121	0.1262
53	Device	103	0.2953
54	Drug delivery	18	1.688
55	Drug delivery system	12	2.2359
56	Effect	39	1.566
57	Efficiency	36	1.3543
58	Electronic	31	0.42
59	Emphasis	15	0.3194
60	End	14	3.4365
61	Energy	21	0.3203
62	Fabrication	94	0.0839
63	FDM	17	2.1511
64	Field	139	0.2792
65	Flexibility	25	0.1635
66	Focus	25	0.1574
67	Functionality	19	0.5147
68	Future	15	0.5737
69	Future perspective	13	0.814
70	Future research	11	0.4278
71	Gap	17	0.4915
72	Healthcare	13	3.8962
73	High strength	12	0.7721
74	Implant	55	1.7826
75	Industry	73	0.1544
76	Influence	13	0.9519
77	Innovation	21	0.8288
78	Insight	24	0.1684
79	Integration	47	0.2946
80	Issue	35	0.7683
81	Lack	12	0.6872
82	Laser	22	0.648
83	Laser powder bed fusion	17	3.1108
84	Lattice structure	18	2.7009
85	Layer	37	0.2838
86	Limitation	47	0.2174

Table 6.1 (continued)

ID	Term	Occurrences	Relevance score
87	LPBF	19	2.9668
88	Machine learning	11	0.5384
89	Magnetic material	13	3.1208
90	Manufacturing	77	0.1059
91	Material extrusion	11	0.4756
92	Mechanical performance	16	3.2542
93	Mechanical property	67	1.0358
94	Mechanism	43	0.4035
95	Medicine	34	2.4817
96	Metal	41	0.3696
97	Metamaterial	19	0.7839
98	Microstructure	36	0.6414
99	MPA	17	5.0743
100	Multimaterial	10	1.983
101	Nanocomposite	31	0.7141
102	Need	32	0.2964
103	Number	15	0.4929
104	Opportunity	42	0.2683
105	Organ	13	2.6881
106	Overview	32	0.1872
107	PA12	26	9.1946
108	Paper	60	0.1125
109	Parameter	52	0.4553
110	Part	87	1.023
111	Particle	34	1.9246
112	Patient	14	2.9976
113	PBF	13	3.771
114	Performance	58	0.4311
115	Perspective	28	0.1983
116	Polyamide	18	8.3458
117	Polymer	128	0.272
118	Polymer composite	17	1.6207
119	Polymeric composite	10	0.8565
120	Possibility	18	1.3618
121	Potential	41	0.4945
122	Potential application	16	0.9296
123	Powder	111	2.1276
124	Powder bed fusion	24	1.7113
125	Precision	23	0.5317
126	Preparation	21	1.1367
127	Printability	17	0.4842
128	Printing	87	0.2148
129	Process	164	0.1441
130	Process parameter	15	0.7033
131	Product	40	0.1225
132	Production	56	0.1593
133	Progress	42	0.7487

Table 6.1 (continued)

ID	Term	Occurrences	Relevance score
134	Property	144	0.353
135	Prospect	23	0.4378
136	Quality	35	1.5233
137	Rapid prototyping	14	0.3553
138	Reader	14	1.2015
139	Recent advance	30	0.3962
140	Recent advancement	11	0.51
141	Recent year	12	0.5765
142	Research	61	0.0525
143	Researcher	27	0.8558
144	Review	160	0.1806
145	Robotic	18	1.9608
146	Role	36	0.5047
147	Scaffold	47	1.0037
148	Selective laser	11	0.6065
149	Selective laser sintering	49	0.9996
150	Sensor	26	0.4634
151	Shape	30	0.5471
152	SiC ceramic	11	2.9029
153	Size	27	1.3281
154	SLA	15	0.9171
155	SLS	34	0.8372
156	Smart material	16	2.1321
157	Soft material	16	1.4996
158	Speed	22	0.3703
159	State	24	0.4091
160	Stereolithography	21	0.649
161	Strategy	52	0.1673
162	Strength	51	2.01
163	Structure	146	0.2712
164	Study	103	0.3395
165	Sustainability	16	0.9123
166	System	85	0.2752
167	Technique	156	0.1008
168	Technology	247	0.4833
169	Tensile strength	12	5.5879
170	Term	21	0.3145
171	Time	30	0.3154
172	Tissue engineering	28	1.4719
173	Type	43	0.4041
174	Use	55	0.5485
175	Variety	14	0.2697
176	Vat photopolymerization	27	1.3897
177	Way	18	0.4871
178	Wide range	15	0.5656
179	Work	45	0.3377

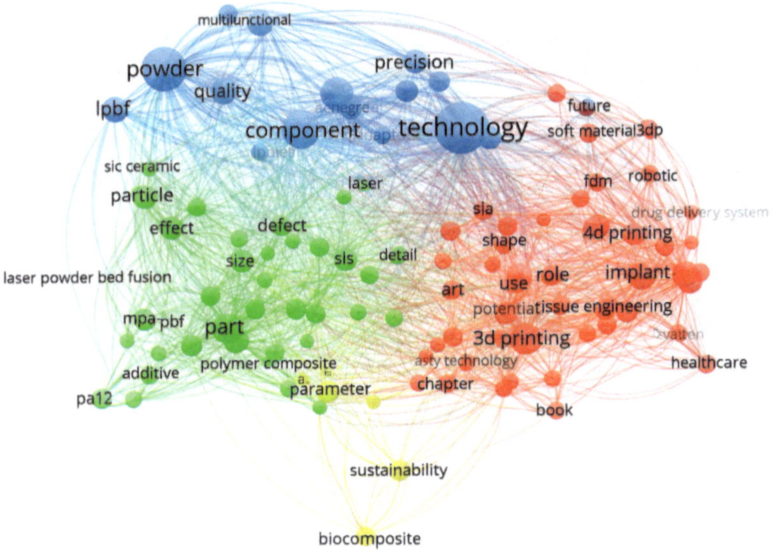

Figure 6.1: Network diagram based on the specific keyword "selective laser sintering and powder bed fusion piezoelectric materials printing".

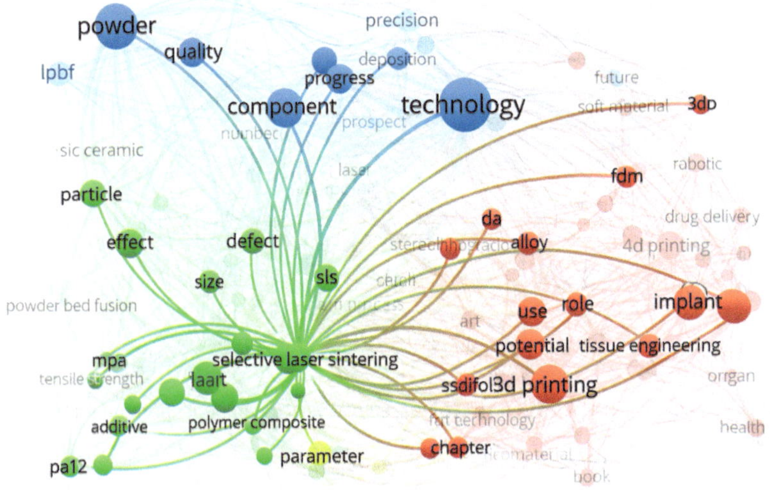

Figure 6.2: Utilization of SLS 3D printing across various fields, including biomedical applications.

In the present study, the CAD model of the compressive sample (ASTM E8) was developed in SOLIDWORKS and converted to a printable format (.stl) for use with the 3DXpert software on the metal printer. Then, using the metal printer software, the ". Fab format file" was generated for sample fabrication. Figure 6.6a shows the CAD

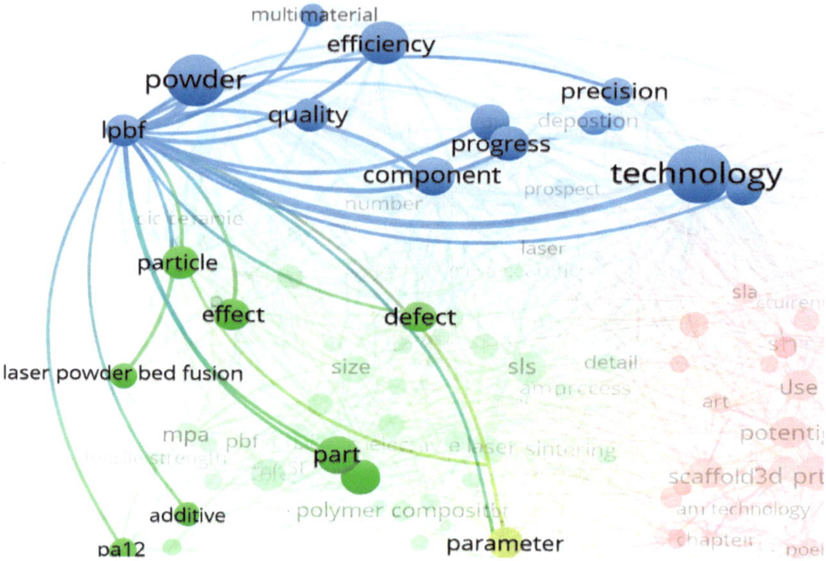

Figure 6.3: Utilization of LPBF 3D printing across various fields, including biomedical applications.

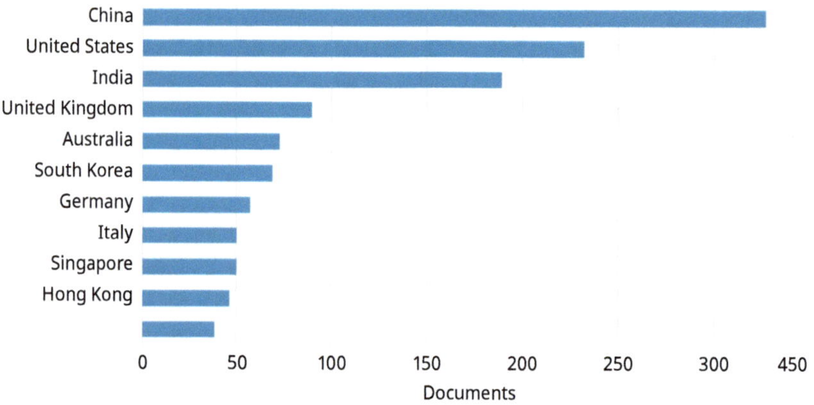

Figure 6.4: Country-wise work reported on SLS and LPBF-based piezoelectric material printing for biomedical applications.

model of the compressive sample. Before fabricating compressive samples, the 17-4PH SS powder with a particle size of ~30 µm was sieved at ~80 Hz and heated in a muffle furnace to remove moisture and prevent cracks and failures during fabrication. For the fabrication of a compressive sample, the LPBF setup with 17-4PH SS material was used, with printing parameters including a laser power of 120 W, three different meta-structures (octet-truss, honeycomb, and diamond), and a scanning speed of 1,200 mm/s. The samples were fabricated so that the top and bottom were solid parts,

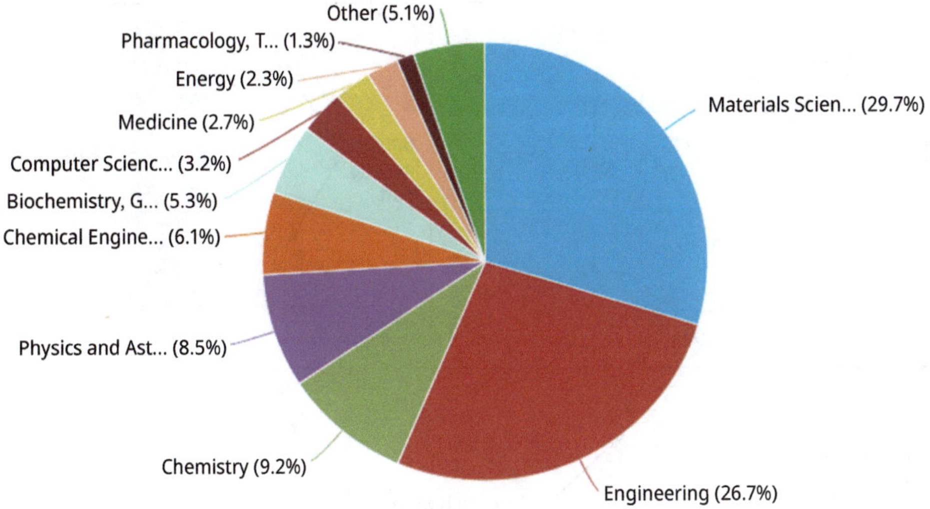

Figure 6.5: Area-wise work reported on SLS and LPBF-based piezoelectric material printing for biomedical applications.

and the intermediate zone had three different octet-truss, honeycomb, and diamond meta-structures. Figure 6.6b–d shows the three different meta-structures used to fabricate compressive samples. Further, a finite element analysis was performed to assess boundary distortion and residual stress using the 3D Xpert software to avoid structural failure and wrapping effects, and to minimize the residual stress and boundary distortion in the overall fabricated samples. Figure 6.6e and f shows the PBF-LB setup and fabricated compressive samples. After sample fabrication, wire-cut electric discharge machining was used to remove the compressive samples from the build plate of the metal printer. Further, a 50 kN universal testing machine (UTM) was used to test the compressive samples at a feed rate of 0.5 mm/min and compress the coupons to 30% of their length. Figure 6.6g shows the samples being tested on the UTM. After the UTM test, the fractured samples were analyzed using scanning electron microscopy (SEM), energy-dispersive spectroscopy (EDS), porosity %, and average grain size distribution. Furthermore, the Thermo-Calc software (using the EDS data) in static mode was used to analyze the phase transformations and thermodynamic behavior of the samples.

6.3 Results and discussion

Figure 6.7 and Table 6.2 show the observed compressive force-position output of the fabricated samples. From Table 6.2, it was observed that all three samples have a high

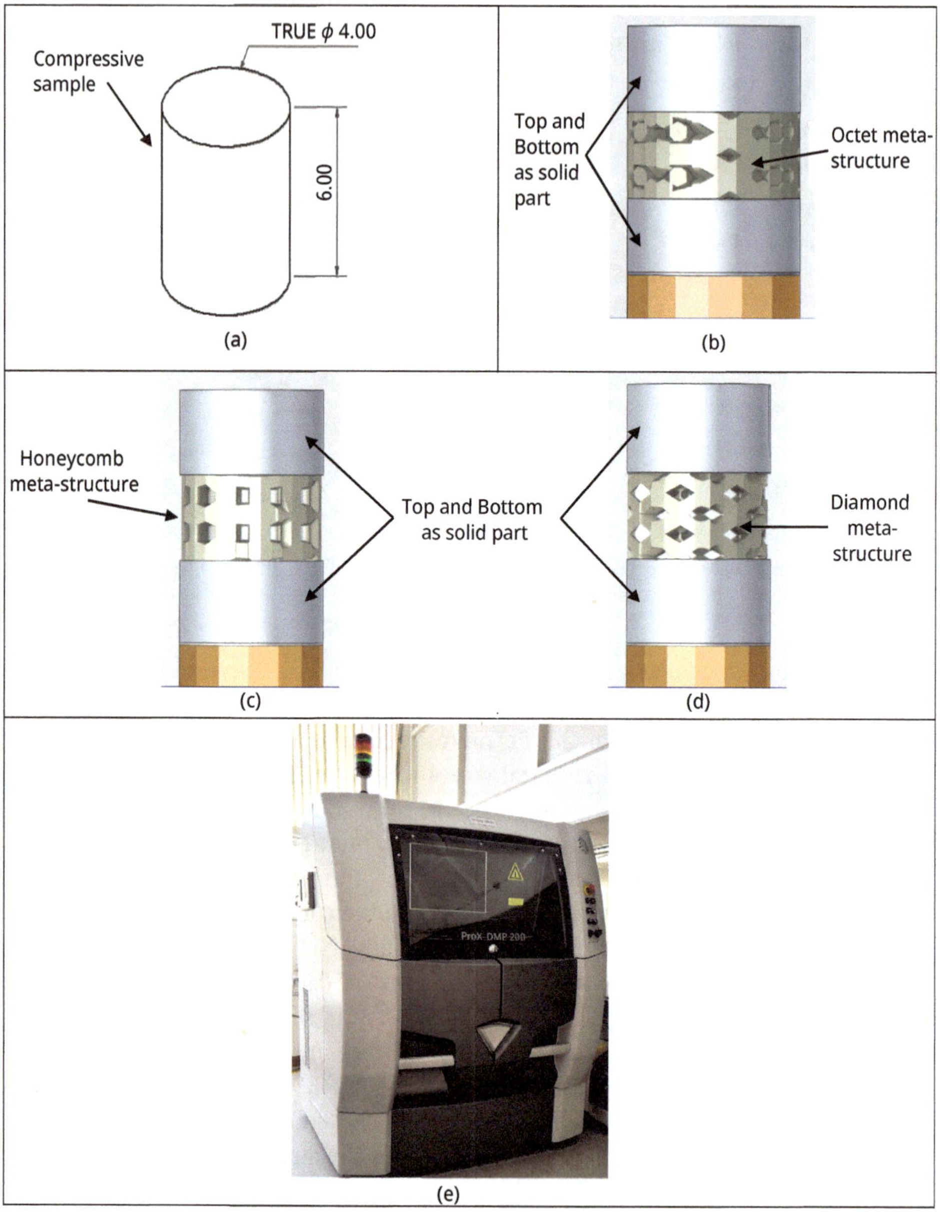

Figure 6.6: (a) CAD model of compressive sample, (b) sample with octet-truss, (c) honeycomb, (d) diamond as meta-structure, (e) PBF-LB setup, (f) fabricated samples, and (g) testing of compressive sample.

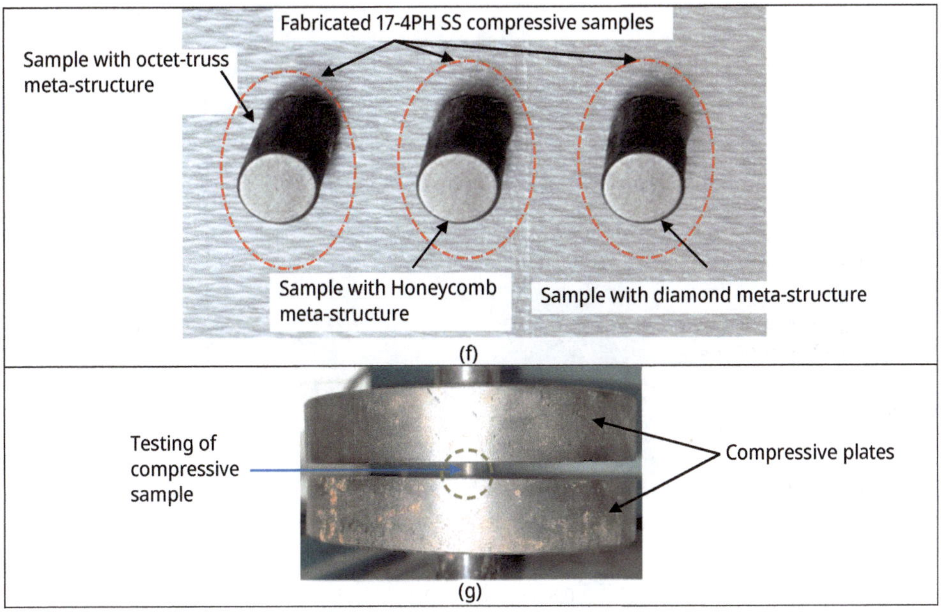

Figure 6.6 (continued)

load-bearing capacity, as evidenced by their compressive strengths: 6,530, 5,155, and 6,610 N, respectively. All these results lead to the conclusion that the meta-structures produced thus far exhibit good mechanical strength for applications in fields where load-sharing with bone is anticipated, such as orthopedics or biomedical applications. Compared with the existing literature, the mean compressive force of ~7,169 N between the tibia and the implant showed good agreement with the results. The sample with an octet-truss as the meta-structure showed the highest compressive strength among the three samples and was also more similar to the previously reported bone-implant condition, indicating that this meta-structure is likely to be optimal, as it provides a suitable carrying load capacity for implant applications, offers better performance, and provides more accurate and reliable mechanical interaction with postoperative tibia behavior.

Table 6.2: Observed compressive force-stress of fabricated samples.

Output parameters		
Compressive samples	**Compressive force (MPa)**	**Compressive stress (MPa)**
Sample with octet-truss meta-structure	6,610	526
Sample with diamond meta-structure	6,530	520
Sample with honeycomb meta-structure	5,160	410

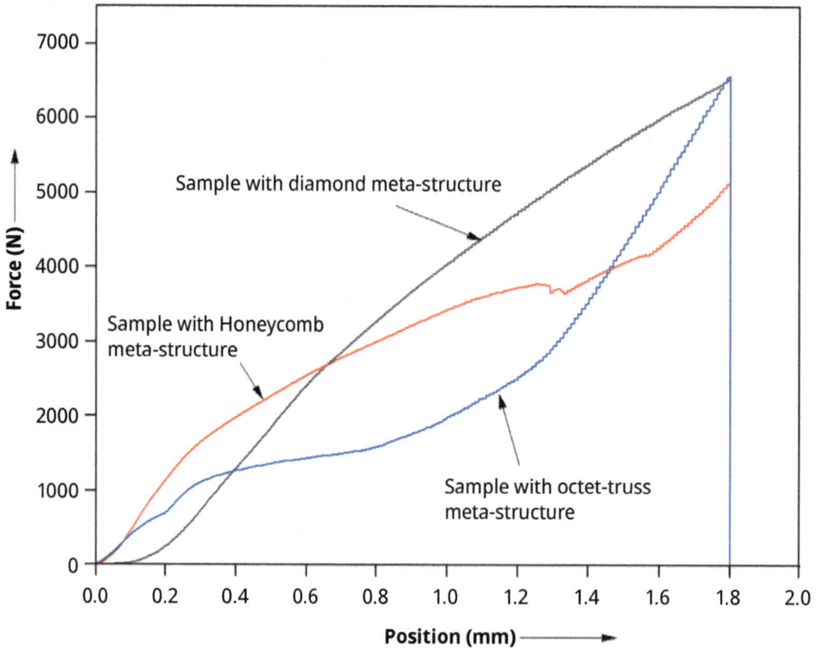

Figure 6.7: Observed force versus position of three compressive samples.

Figure 6.8 shows the SEM, EDS, porosity%, and average grain size of the three fabricated samples. Based on Table 6.2, the sample with the octet-truss meta-structure had the highest compressive strength and was considered the best-performing design in terms of mechanical performance. However, higher porosity is preferred for orthopedic and biological applications because it promotes bone ingrowth, vascularization, and improved osseointegration. On the other side, increased porosity leads to decreased strength; therefore, there must be an optimal balance between mechanical properties and biological function. The sample with the honeycomb meta-structure has the highest porosity (~19.11%), making it better for biological integration while maintaining sufficient mechanical strength. Thus, the honeycomb meta-structure demonstrates an idealized balance between mechanical performance and biological compatibility and may be considered the best meta-structure design for implant applications.

Figure 6.9a shows the melt pool analysis of fabricated samples. The melt pool analysis of the manufactured samples at 120 W laser power and 1,200 mm/s scanning speed indicates a steady controlled melting state, sufficient melt pool depth, and a localized high-temperature zone beneath the laser spot. That means enough energy was provided to achieve appropriate melting and coating-layer interbonding but without excessive heat damage. This steady melt-pool behavior led to better densification, which improved metallurgical bonding and reduced lack-of-fusion defects, thereby yielding the desired com-

Observations	SEM	EDS	Porosity %	AGSN
Sample with octet-truss meta-structure	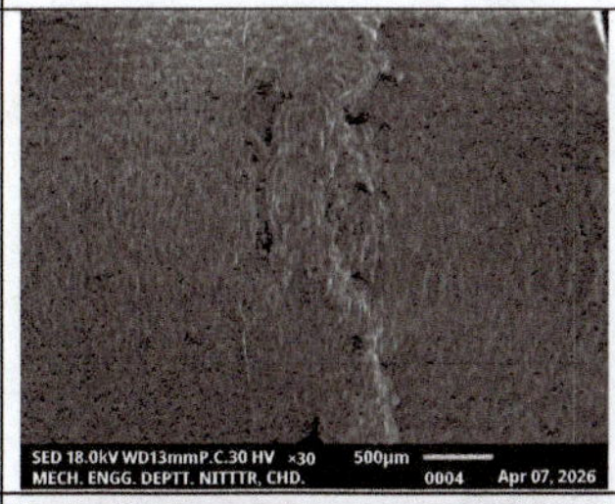	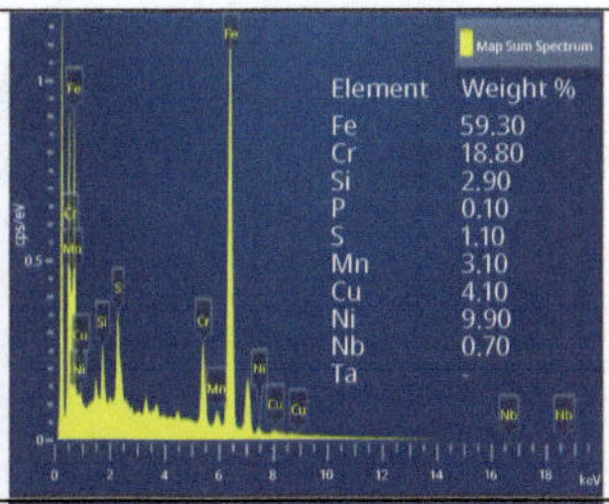	13.77	0.5
Sample with diamond meta-structure	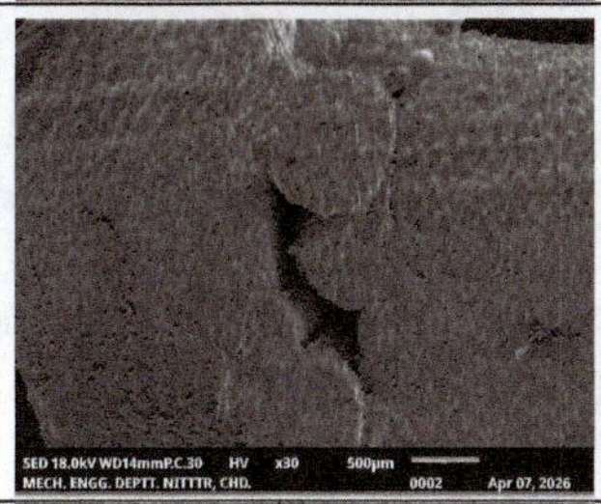	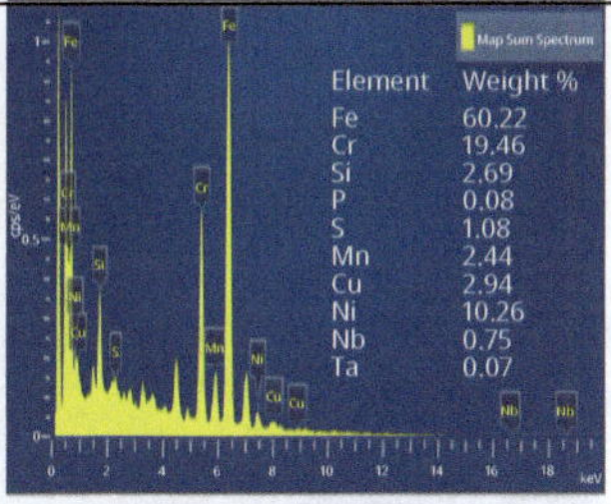	8.03	1
Sample with honeycomb meta-structure	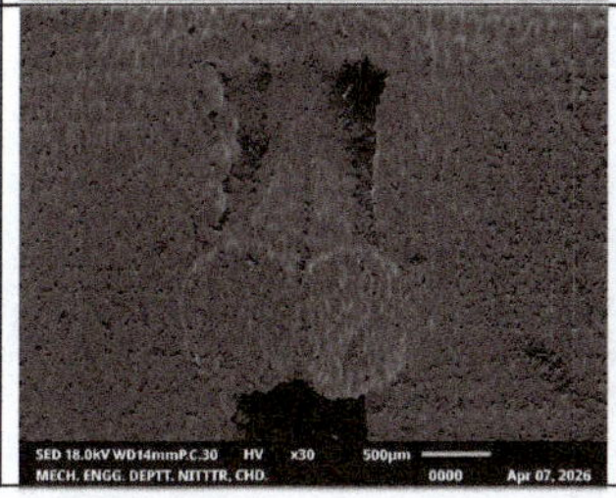	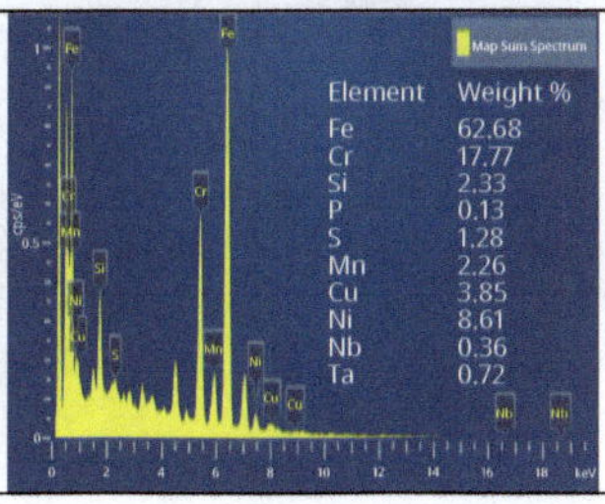	19.11	1

Figure 6.8: SEM, EDS, porosity%, and average grain size of the three fabricated samples.

pressive properties of the produced samples. As a result, the beneficial melt-pool characteristics observed during this processing regime can be attributed to the samples' remarkable compressive performance. Figure 6.9b shows the temperature-distance analysis of fabricated samples. The temperature increases continuously from the substrate region into the melt pool, with a maximum above the liquidus temperature (2,100 K) and approaching the evaporation-limited range (2,800–3,000 K). The temperature-distance analysis shows a rapid decrease, indicating rapid solidification followed by a gradual drop, with a controlled thermal profile that promotes strong metallurgical bonding between layers and the formation of a fine microstructure, suggesting a proper molten state and solidification with a high temperature gradient. Adequate and homogeneous melting, without excessive overheating, results in fewer defects, such as keyhole effects, porosity, and lack of fusion, which may explain the better compressive properties (such as strength) observed in fabricated samples.

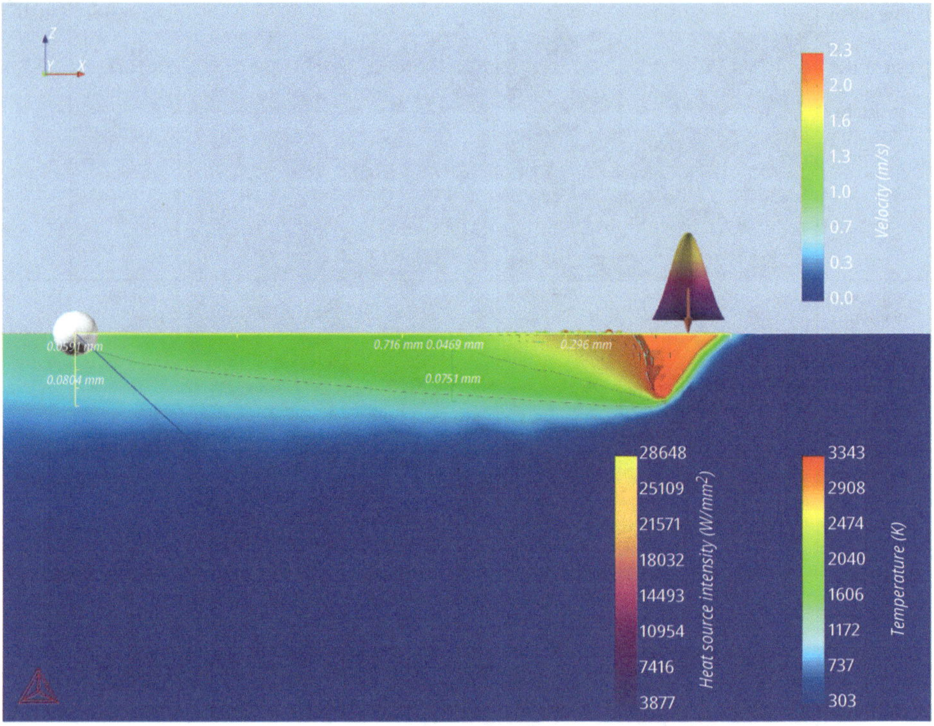

(a)

Figure 6.9: Melt pool and temperature versus distance analysis of fabricated samples.

Figure 6.10 shows the thermal gradient versus solidification analysis of fabricated samples. From the analysis of temperature gradients and solidification rates, there is a negative correlation between large thermal gradients and low solidification rates, and vice versa. Compared with lower thermal gradients, higher thermal gradients promote rapid heat extraction and directional solidification, thereby favoring a finer microstructure and better grain refinement. Moderate solidification rates also allow for consistent phase development without excessive heat stress, and the increased interlayer adhesion and reduced defects/enhanced microstructural quality induced by this high-temperature gradient, alongside a controlled solidification rate, are responsible for the superior compressive properties observed in the fabricated samples.

Figure 6.11 and Table 6.3 show the phase-volume fraction static equilibrium analysis of the volume-phase fracture-temperature for the fabricated samples. The main phases constituting the microstructure are shown in the phase analysis of the produced samples, BCC_B2 (0.578) and BCC_A2 (0.201), resulting in a strong and stable matrix and

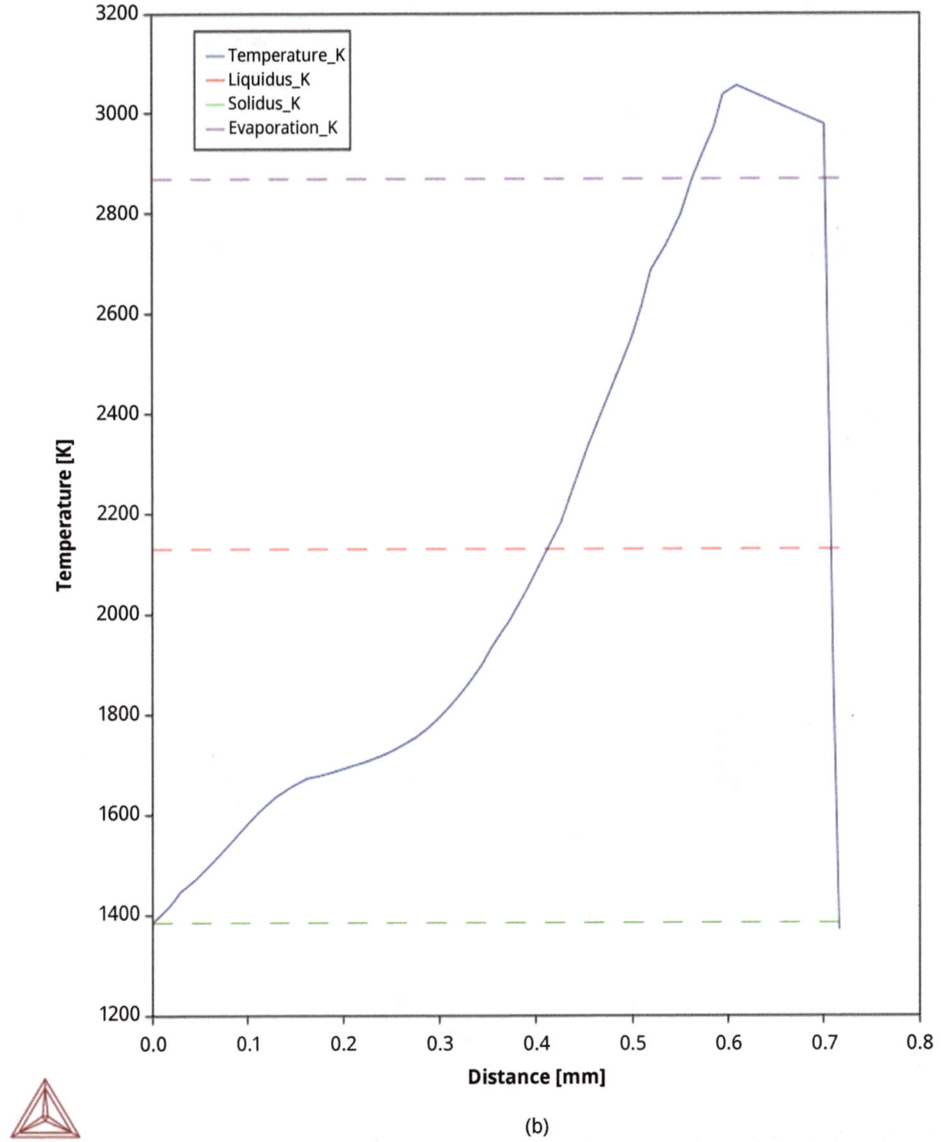

(b)

Figure 6.9 (continued)

small fractions, additional amounts of secondary phases, such as the G phase, M3P, and martensitic phase (MS_B1), are present, whilst some trace FCC phases, such as FCC_A1 and FCC_L12, are also observed. Although secondary and FCC phases can enhance toughness and microstructural stability, the primary BCC phases lead to higher strength. This balanced phase composition correlates well with the compressive properties measured

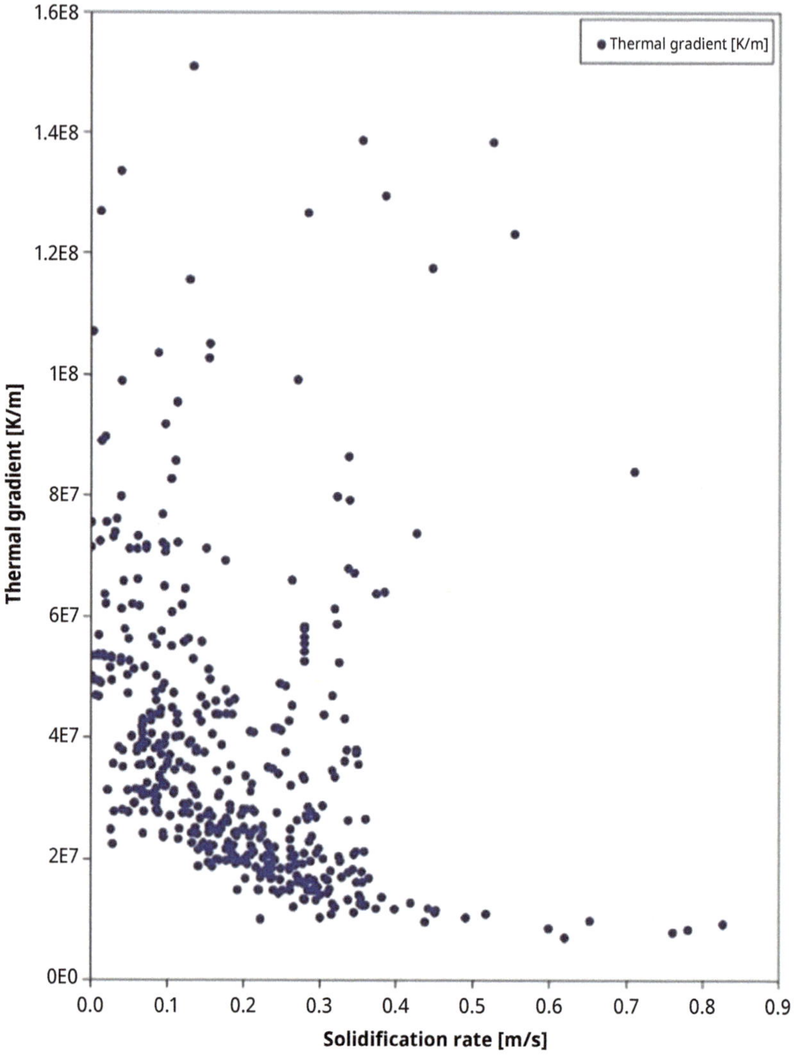

Figure 6.10: Thermal gradient versus solidification analysis of fabricated samples.

for manufactured samples and supports good mechanical performance. The phase analysis shows that the produced samples are largely composed of BCC and FCC phases, which are inherently centrosymmetric and therefore lack intrinsic piezoelectricity. However, induced martensite and B2 structures are not piezoelectric phases, though they may contribute to strain-induced or pseudo-electromechanical responses. Hence, the existing material system is more suitable for structural and load-bearing biomedical applications than for direct piezoelectric applications.

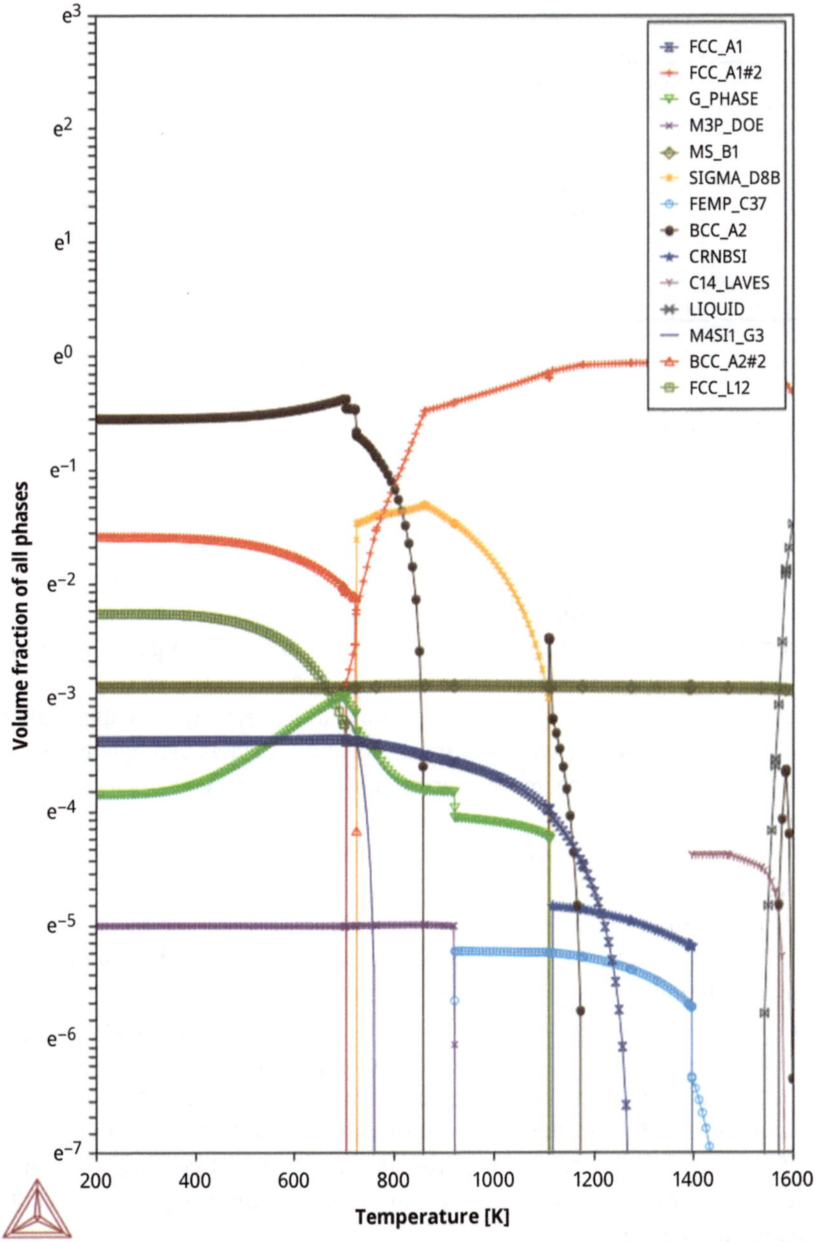

Figure 6.11: Volume fraction of all phases versus temperature of fabricated samples.

Table 6.3: Observed stable phases at room temperature.

Phases	Volume fraction
BCC_A2#1	0.201
BCC_B2#1	0.578
FCC_A1#1	0.034
FCC_L12#1	0.104
G_PHASE#1	0.022
M3P_D0E#1	0.007
MS_B1#1	0.055

6.4 Conclusions

The study highlights the fabrication of a compressive sample based on three meta-structures (octet-truss, honeycomb, and diamond) of 17-4PH SS fabricated via PBF-LB to evaluate the piezoelectric behavior for biomedical and implant applications. The following are the conclusions from this study:

- The observed compressive properties indicate that all three samples have a high load-bearing capacity, as evidenced by their compressive strengths of 6,530, 5,155, and 6,610 N, respectively, and exhibit good mechanical strength for applications in fields where load-sharing with bone is anticipated, such as orthopedics or biomedical applications. Furthermore, the sample with an octet-truss as the meta-structure showed the highest compressive strength, close to that of the canine tibia, likely optimal, as it provides a suitable carrying load capacity for implant applications.
- In terms of porosity (%), the honeycomb meta-structured sample has the highest porosity (~19.11%), making it better for biological integration while maintaining sufficient mechanical strength. Further, demonstrates an idealized balance between mechanical performance and biological compatibility and may be considered the best meta-structure design for implant applications.
- The melt pool analysis indicates a steady, controlled melting state, a sufficient melt pool depth, and a localized high-temperature zone beneath the laser spot, with sufficient energy to achieve appropriate melting and coating-layer interbonding without excessive heat damage, attributed to the samples' remarkable compressive performance.
- In the thermal gradient-solidification analysis, moderate solidification rates allow for consistent phase development without excessive heat stress, and the increased interlayer adhesion and reduced defects/enhanced microstructural quality induced by this high-temperature gradient, alongside a controlled solidification rate, are responsible for the superior compressive properties in the samples.

– In the phase fraction-temperature analysis, the main phases constituting the microstructure are shown in the phase analysis in suitable samples, resulting in a strong and stable matrix and small fractions, additional amounts of secondary phases (such as the G phase, M3P, and martensitic phase), are present, while some trace FCC phases, such as FCC_A1 and FCC_L12, are also observed. This balanced phase composition correlates well with the compressive properties measured for manufactured samples and supports good mechanical performance.

References

[1] Choudhury D, Ponneganti S, Radhakrishnanand P, Murty US, Banerjee S. Selective laser sintering additive manufacturing of solid oral dosage form: effect of laser power and hatch spacing on the physico-technical behaviour of sintered printlets. Applied Materials Today. 2023 Dec 1;35:101943.
[2] Fina F, Goyanes A, Gaisford S, Basit AW. Selective laser sintering (SLS) 3D printing of medicines. International journal of pharmaceutics. 2017 Aug 30;529(1–2):285–93.
[3] Awad A, Fina F, Goyanes A, Gaisford S, Basit AW. 3D printing: Principles and pharmaceutical applications of selective laser sintering. International Journal of Pharmaceutics. 2020 Aug 30;586:119594.
[4] Gueche YA, Sanchez-Ballester NM, Cailleaux S, Bataille B, Soulairol I. Selective laser sintering (SLS), a new chapter in the production of solid oral forms (SOFs) by 3D printing. Pharmaceutics. 2021 Aug 6;13(8):1212.
[5] Shammout H, Saker R, Ludasi K, Sovány T. An overview of selective laser sintering technology: principles, formulations, applications and promising potentials in pharmaceutics. Journal of Drug Delivery Science and Technology. 2025 Sep 1;111:107166.
[6] Song Y, Ghafari Y, Asefnejad A, Toghraie D. An overview of selective laser sintering 3D printing technology for biomedical and sports device applications: Processes, materials, and applications. Optics & Laser Technology. 2024 Apr 1;171:110459.
[7] Du Y, Liu H, Shuang J, Wang J, Ma J, Zhang S. Microsphere-based selective laser sintering for building macroporous bone scaffolds with controlled microstructure and excellent biocompatibility. Colloids and Surfaces B: Biointerfaces. 2015 Nov 1;135:81–89.
[8] Rahmani R, Lopes SI, Prashanth KG. Selective laser melting and spark plasma sintering: a perspective on functional biomaterials. Journal of Functional Biomaterials. 2023 Oct 16;14(10):521.
[9] Aho J, Bøtker JP, Genina N, Edinger M, Arnfast L, Rantanen J. Roadmap to 3D-printed oral pharmaceutical dosage forms: feedstock filament properties and characterization for fused deposition modeling. Journal of Pharmaceutical Sciences. 2019 Jan 1;108(1):26–35.
[10] Liu J, Zhou Y, Lu J, Cai R, Zhao T, Chen Y, Zhang M, Lu X, Chen Y. Injectable, tough and adhesive zwitterionic hydrogels for 3D-printed wearable strain sensors. Chemical Engineering Journal. 2023 Nov 1;475:146340.
[11] Olakanmi EO. Selective laser sintering/melting (SLS/SLM) of pure Al, Al–Mg, and Al–Si powders: Effect of processing conditions and powder properties. Journal of Materials Processing Technology. 2013 Aug 1;213(8):1387–405.
[12] Singh S, Kaur D, Singh M, Balu R, Mehta A, Vasudev H. Challenges and issues in manufacturing of components using polymer-based selective laser sintering (SLS): a review. International Journal on Interactive Design and Manufacturing (IJIDeM. 2024 Aug;8:1–24.

[13] Kruth JP, Badrossamay M, Yasa E, Deckers J, Thijs L, Van Humbeeck J, Zhao W, Ye J, Zhu D. Part and material properties in selective laser melting of metals. In Proceedings of the 16th International Symposium on Electromachining (ISEM XVI) 2010 Apr 1 (pp. 3–14). Shanghai Jiao Tong Univ Press.

[14] Kushwaha AK, Rahman MH, Slater E, Patel R, Evangelista C, Austin E, Tompkins E, McCarroll A, Rajak DK, Menezes PL. Powder bed fusion-based additive manufacturing: SLS, SLM, SHS, and DMLS. In Tribology of additively manufactured materials 2022 Jan 1 (pp. 1–37). Elsevier.

[15] Gu D. Materials creation adds new dimensions to 3D printing. Science Bulletin. 2016 Nov;61 (22):1718–22.

[16] Gao B, Zhao H, Peng L, Sun Z. A review of research progress in selective laser melting (SLM). Micromachines. 2022 Dec 25;14(1):57.

[17] Aufa AN, Hassan MZ, Ismail Z, Ramlie F, Jamaludin KR, Daud MY, Ren J. Current trends in additive manufacturing of selective laser melting for biomedical implant applications. Journal of Materials Research and Technology. 2024 Jul 1;31:213–43.

[18] Joshua RJ, Raj SA, Hameed Sultan MT, Łukaszewicz A, Józwik J, Oksiuta Z, Dziedzic K, Tofil A, Shahar FS. Powder bed fusion 3D printing in precision manufacturing for biomedical applications: a comprehensive review. Materials. 2024 Feb 5;17(3):769.

[19] Gatto ML, Cerqueni G, Groppo R, Santecchia E, Tognoli E, Defanti S, Mattioli-Belmonte M, Mengucci P. Improved biomechanical behavior of 316L graded scaffolds for bone tissue regeneration produced by laser powder bed fusion. journal of the mechanical behavior of biomedical materials. 2023 Aug 1;144:105989.

[20] Jumadi J, Shaari MS, Fauadi MH, Kumaresan R, Premono A, Saxena KK, Kadirgama K, Harun WS. Selective laser melting of stainless steel for biomedical implants: A critical review of processing, microstructure, and performance. The International Journal of Advanced Manufacturing Technology. 2026 Feb;9:1–24.

[21] Marin E, Lanzutti A. Biomedical applications of titanium alloys: a comprehensive review. Materials. 2023 Dec 25;17(1):114.

[22] Hassan MZ, Aufa AN, Brytan Z, Tanski T, Ren J, Mazurkiewicz J, Sani MS, Ismail Z. Selective laser melting of magnesium alloys for biomedical applications: Microstructure to clinical translation. Journal of Materials Research and Technology. 2025 Sep 19.

[23] Gracheva A, Polozov I, Popovich A. Additive manufacturing of biodegradable metallic implants by selective laser melting: current research status and application perspectives. Metals. 2025 Jul 4;15(7):754.

[24] Li Z, Lu M, Lei H, Zou S, Liu R, Wang Y, Long J, Zhang B, You Q, Luo Y, Zhou Y. Selective laser melting fabrication of body-temperature shape memory NiTi alloy for 4D-printed orthopedic implants. Virtual and Physical Prototyping. 2024 Dec 31;19(1):e2413181.

[25] Ginestra P, Ferraro RM, Zohar-Hauber K, Abeni A, Giliani S, Ceretti E. Selective laser melting and electron beam melting of Ti6Al4V for orthopedic applications: a comparative study on the applied building direction. Materials. 2020 Dec 7;13(23):5584.

[26] Puebla K, Murr LE, Gaytan SM, Martinez E, Medina F, Wicker RB. Effect of melt scan rate on microstructure and macrostructure for electron beam melting of Ti-6Al-4V. Materials Science and Applications. 2012 May;3(5):259–64.

[27] Traini T, Mangano C, Sammons RL, Mangano F, Macchi A, Piattelli A. Direct laser metal sintering as a new approach to fabrication of an isoelastic functionally graded material for manufacture of porous titanium dental implants. Dental materials. 2008 Nov 1;24(11):1525–33.

[28] Van Noort R. The future of dental devices is digital. Dental materials. 2012 Jan 1;28(1):3–12.

[29] Vandenbroucke B, Kruth JP. Selective laser melting of biocompatible metals for rapid manufacturing of medical parts. Rapid Prototyping Journal. 2007 Aug 7;13(4):196–203.

[30] Mazzoli A. Selective laser sintering in biomedical engineering. Medical & biological engineering & computing. 2013 Mar;51(3):245–56.

[31] Hrabe N, Quinn T. Effects of processing on microstructure and mechanical properties of a titanium alloy (Ti–6Al–4V) fabricated using electron beam melting (EBM), Part 2: Energy input, orientation, and location. Materials Science and Engineering: A. 2013 Jun 20;573:271–77.

[32] Li Y, Zhou K, Tor SB, Chua CK, Leong KF. Heat transfer and phase transition in the selective laser melting process. International Journal of Heat and Mass Transfer. 2017 May 1;108:2408–16.

[33] Sun J, Yang Y, Wang D. Parametric optimization of selective laser melting for forming Ti6Al4V samples by Taguchi method. Optics & Laser Technology. 2013 Jul 1;49:118–24.

[34] Bose S, Vahabzadeh S, Bandyopadhyay A. Bone tissue engineering using 3D printing. Materials today. 2013 Dec 1;16(12):496–504.

[35] Sing SL, An J, Yeong WY, Wiria FE. Laser and electron-beam powder-bed additive manufacturing of metallic implants: A review on processes, materials, and designs. Journal of Orthopaedic Research. 2016 Mar;34(3):369–85.

Bharat Kalia*, Gurwinder Singh, Ravinder Singh

Chapter 7
Hybrid, metastructures, and multimaterial-based 3D printing for piezoelectric functional structures

Abstract: Piezoelectric materials have become increasingly important for sensing and actuation because they can readily convert mechanical energy into electrical signals. Conventional piezoceramics, such as zirconate titanate (PZT), have excellent electromechanical properties but are brittle and require further processing. By comparison, piezoelectric polymers such as polyvinylidene fluoride (PVDF) offer greater flexibility and are easier to manufacture, but with lower functional efficiency. Recent studies have aimed to address this gap by considering architected and metamaterial-based designs, in which deformations and strain amplification in engineered geometries (lattices, trusses, auxetic structures) are employed to optimize the electromechanical response. Simultaneously, with the development of multimaterial additive manufacturing (AM), it is now possible to produce such complex hybrid systems with an extremely high level of precision and design freedom. One of these technologies is laser powder bed fusion (LPBF), which has emerged as particularly promising for enabling the incorporation of piezoelectric elements into metallic structures, through postfabrication integration or by incorporating them directly into the structure during the build process. There are still challenges, such as thermal mismatch, material fragility, and reliable electrical interfacing, in the design and fabrication of such systems. This study provides a review of hybrid, metastructured, and multimaterial 3D printing methods for piezoelectric functional structures, including major design strategies, integration techniques, and process considerations for LPBF-enabled systems. The selected case studies are discussed, highlighting self-sensing metamaterial structures, in situ integration, and prestressing of piezoelectric stacks, and the application of piezoelectric ultrasonic actuation to control melt pool dynamics during fabrication. Lastly, the chapter presents a pilot study to build a multimaterial auxetic lattice structure using 17-4 PH

***Corresponding author: Bharat Kalia,** Mechanical Engineering Department, National Institute of Technical Teachers Training and Research, Chandigarh, India; University Center for Research and Development, Chandigarh University, Mohali, Punjab, India e-mail: bharatkalia98@gmail.com
Gurwinder Singh, Mechanical Engineering Department, National Institute of Technical Teachers Training and Research, Chandigarh, India
Ravinder Singh, Mechanical Engineering Department, Chandigarh Engineering College-CGC, Landran, Mohali, India
Rupinder Singh, Mechanical Engineering Department, National Institute of Technical Teachers Training and Research, Chandigarh, India

 | https://doi.org/10.1515/9783112218846-007

stainless steel, Ni-625, and Ti-6Al-4V alloy powders, demonstrating the feasibility of embedded piezoelectric functionality.

Keywords: LPBF, piezoelectric functional structures, metastructures, multimaterial 3D printing, piezoelectric actuation

7.1 Introduction

Mechanical metamaterials have transformed the design of structural systems, in which the base material no longer defines the system's performance but rather its internal architecture [1, 2]. Such engineered materials can exhibit properties rarely observed in naturally occurring systems, owing to carefully designed unit-cell topologies [3, 4]. Such architected or metastructured designs, including lattices and trusses, as well as periodic cellular structures, have recently attracted significant interest, especially with the development of additive manufacturing (AM) technologies that enable them to be fabricated accurately [5, 6]. These advances are particularly applicable to the field of hybrid and multimaterial 3D printing, where geometry and material composition can be jointly controlled to achieve improved functionality [7–9].

The auxetic architectures have become a prominent type of metastructure due to their negative Poisson's ratio, in which the material laterally expands under tension and contracts under compression, unlike traditional materials [10–12]. This unconventional response is due to certain unit cell geometries, including re-entrant honeycombs, rotating geometries, and perforated designs [13–15]. For instance, the inclined struts linked at joints in a concave, bowtie-like geometry form re-entrant and rotating square unit cells [16, 17]. These structures primarily hinge and rotate as they are loaded, rather than simply stretching. This will enable them to distribute strain more evenly and absorb more energy [18, 19]. Due to this controllable deformation behavior, such metastructures are especially applicable in applications that require high impact resistance, flexibility, and efficient load transfer [20–22].

Most metastructures, however, despite their significant mechanical benefits, are unable to detect or respond to environmental changes due to their passive nature, which has imposed limitations on the creation of hybrid functional systems, where structural architectures are coupled with smart materials to add sensing and actuation functions [23]. Piezoelectric materials are of particular importance in this regard, since they permit direct conversion between mechanical and electrical energy, meaning that structures can not only detect external stimuli but also react to them. The inclusion of piezoelectric devices in 3D-printed metastructures via in situ or multimaterial printing is a notable step towards achieving multifunctional systems [24]. This integration naturally leads to the concept of smart structures, where materials design and functionality are all integrated into a single framework. In these systems, parameters such as stress, strain, or vibration are sensed, and actuators embedded allow an adaptive response. These structures can adapt their behavior dynamically to realistic situations when supported by feedback and control strategies, thereby achieving better performance, reliability, and efficiency. Advances in multimaterial AM have also enabled the accurate placement of functional and structural materials within complex geometries, thereby increasing electromechanical coupling and the overall performance of the system.

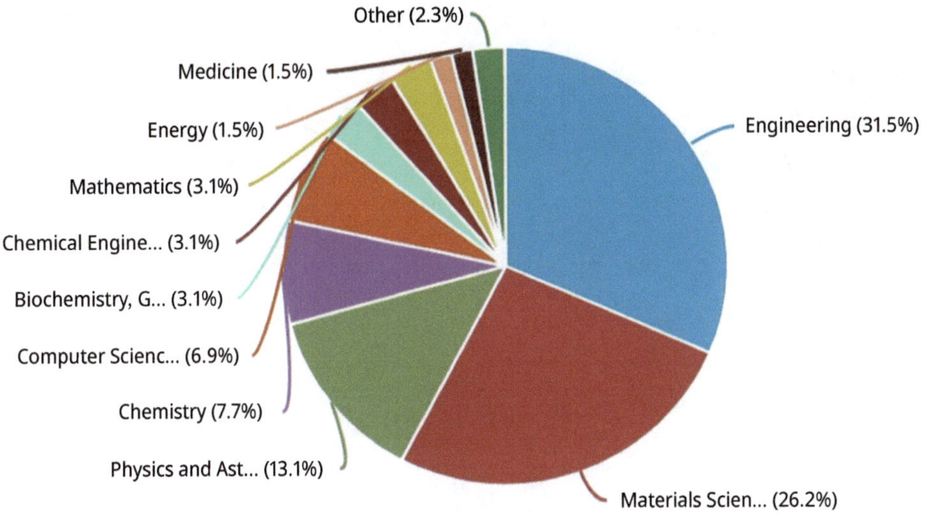

Figure 7.1: Pie chart-based distribution of various types of articles reported on metastructure-based piezoelectric systems.

The bibliometric analysis shows that research on metastructure-based piezoelectric systems has evolved rapidly and is multidisciplinary. The core disciplines in the field are mainly engineering and materials science, which have been significantly contributed to by physics, chemistry, and computational disciplines, indicating high cross-disciplinary integration as indicated in Figure 7.1. In terms of geography, the United States and China are the countries leading research activities as indicated in Figure 7.2.

The time trend indicates an evident increase in publications, particularly after 2018, which implies growing interest in research and expanding technological relevance (Figure 7.3). The distribution also reveals gaps in areas such as multimaterial integration and in situ functionalization in laser powder bed fusion (LPBF). Overall, the trends indicate a field that is growing rapidly, yet with great potential for further innovation and application-driven research. This strategy offers a solid foundation for new applications such as structural health monitoring, biomedical devices, aerospace components, and advanced AM systems.

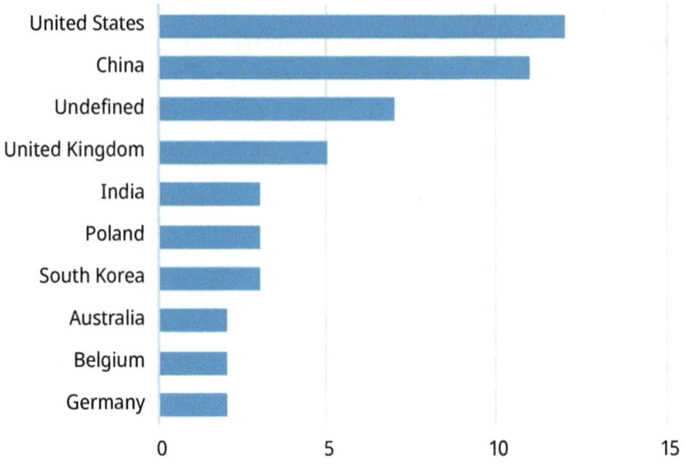

Figure 7.2: Work reported on 3D-printed piezoelectric systems country-wise.

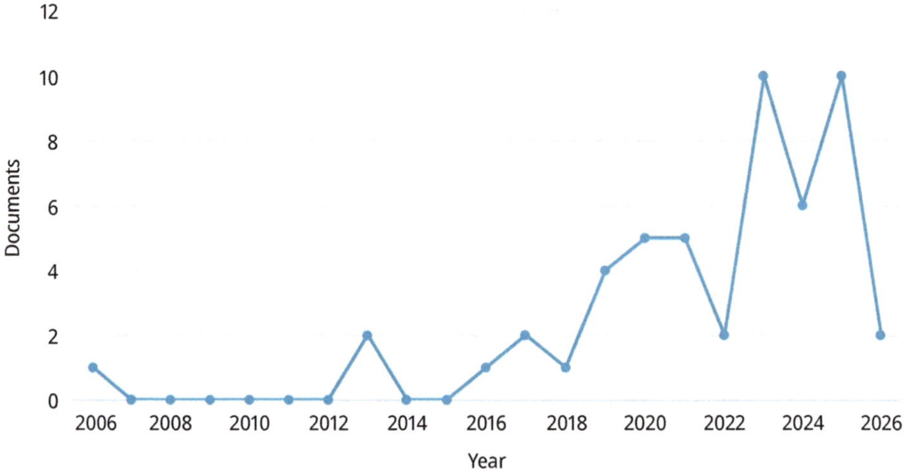

Figure 7.3: Graphical observations for the increasing trend of using 3D printing piezoelectric systems.

7.2 Research gap

According to the literature review, the Scopus database was searched to identify the latest trends in research on 3D-printed metastructure-based piezoelectric systems. The VOSviewer was used to conduct a bibliometric study, which identified 53 articles of interest in this field. Of the 766 extracted keywords, a minimum occurrence of 3 was used to streamline the analysis, resulting in 53 key terms. These words create clear clusters, identifying the interdisciplinary nature of the field and connecting fields such as AM, smart materials, and functional structures, as depicted in Figure 7.4.

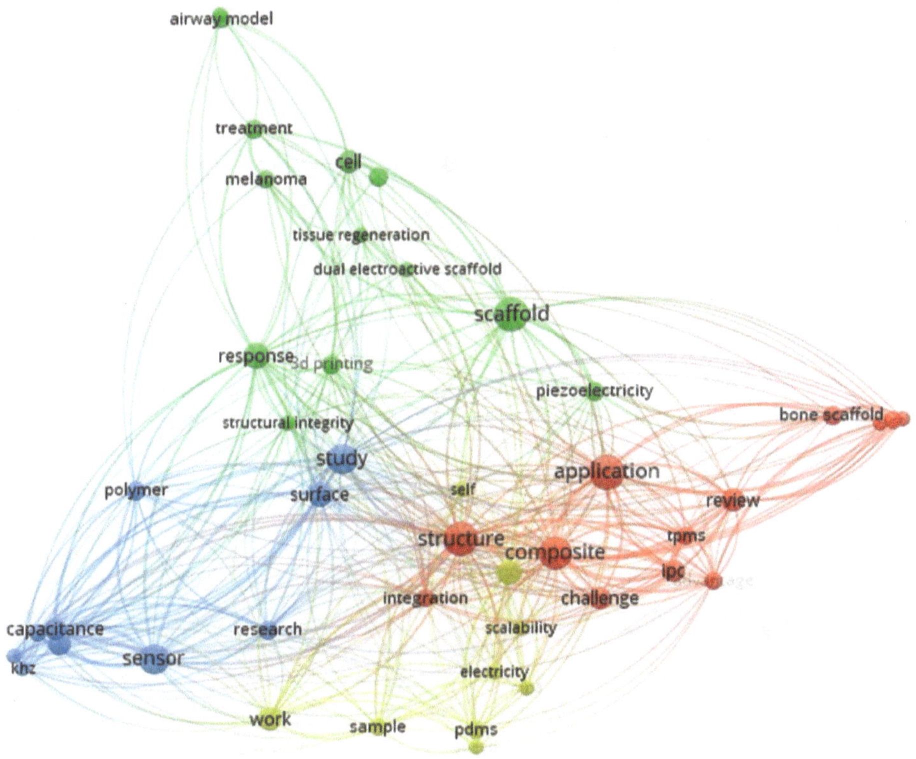

Figure 7.4: Cluster of key terms reported on 3D-printed metastructure-based piezoelectric systems.

The 3D-printing-based piezoelectric sensor fabrication network appears to be highly focused on the selection of materials, processing techniques, and sensing characteristics that reflect the field's development toward the creation of functional devices, as revealed in Figure 7.5.

However, much of the existing work is still limited to single-material systems and traditional design approaches. In comparison, less focus has been on meta-structured designs and multimaterial AM of piezoelectric functional structures. The joint application of architected geometries and material heterogeneity to increase electromechanical performance is a less-explored area, as shown in Figure 7.6.

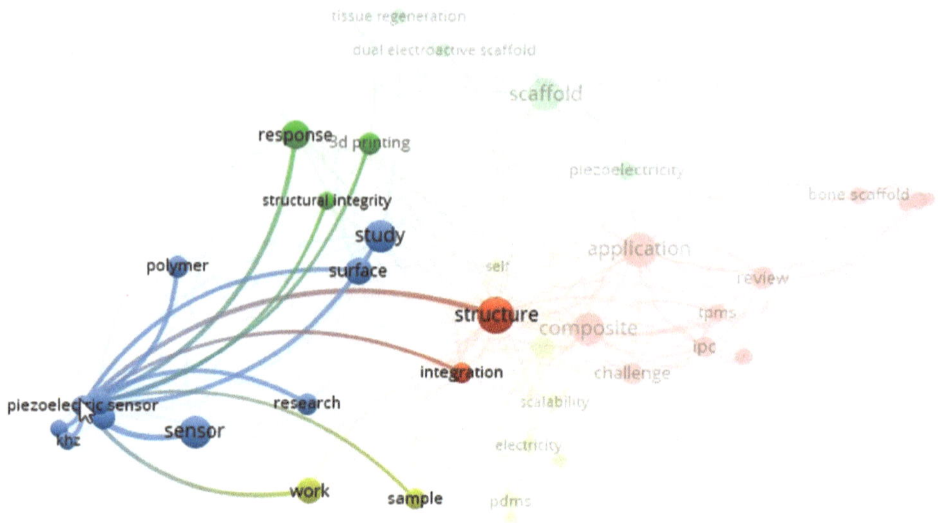

Figure 7.5: Web of terms reported on 3D printing applications for piezoelectric sensor fabrication.

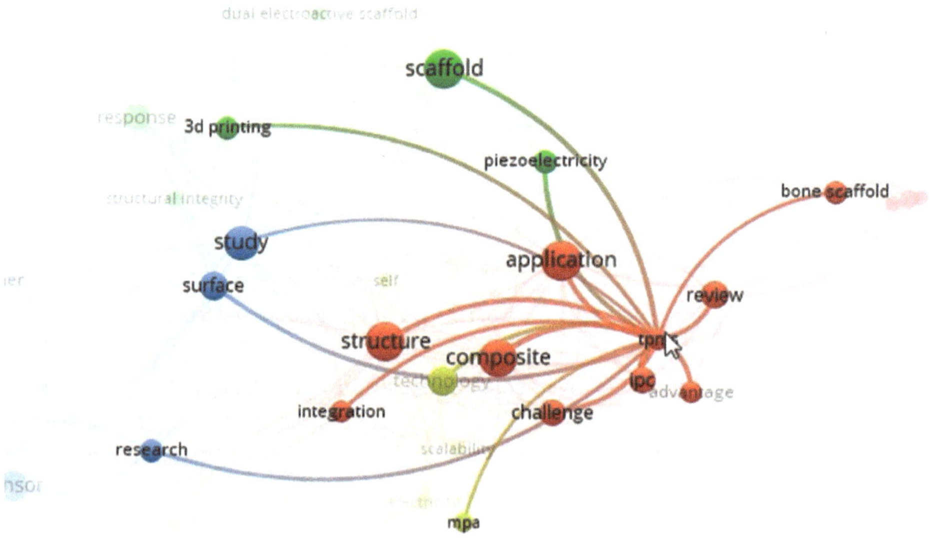

Figure 7.6: Gap in literature related to less reported studies on metastructures and multimaterial-based 3D printing for piezoelectric functional structures.

7.3 Additive manufacturing (LPBF) for piezoelectric functional structures

AM, especially LPBF, has become a groundbreaking pathway towards piezoelectric functional structures with a degree of design freedom and integration unattainable by traditional manufacturing approaches. LPBF is a manufacturing process that involves the selective melting and fusing of metal powder layers layer by layer using a high-energy laser to create complex and precise structures [25, 26]. The process starts by spreading the metal powder evenly on the build platform using a recoater. As per the CAD model, the laser scans a given area and melts the powder to create a solid layer, as depicted in Figure 7.7. After a layer is created, the build platform lowers, and a fresh layer of powder is added [11]. This process is repeated until the part is fabricated. After printing, the component is left to cool in the build chamber for some time, then removed (Figure 7.8). Postprocessing steps, such as heat treatment and surface finishing, may be needed to enhance mechanical properties and achieve the desired surface quality, depending on the application. The layer-based construction of LPBF allows the production of complex metallic structures with high spatial resolution and is thus especially well-suited for the incorporation of piezoelectric elements into load-bearing components.

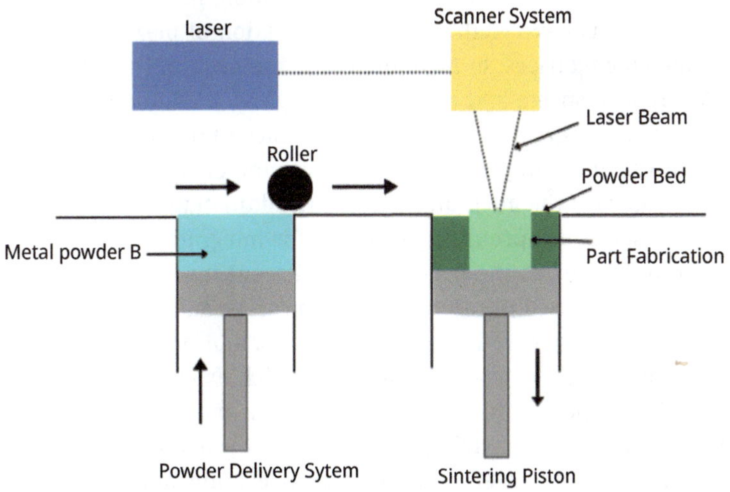

Figure 7.7: Schematic of LPBF process.

This is significant for the creation of next-generation smart structures, where sensing, actuation, and structural capabilities are co-designed rather than incorporated separately. Besides this, the ability to customize internal geometries, including lattice and auxetic structures, enables control of local stress distribution or stress concentration within the component. This can improve the performance of embedded piezoelectric elements by increasing their sensitivity and overall electromechanical response.

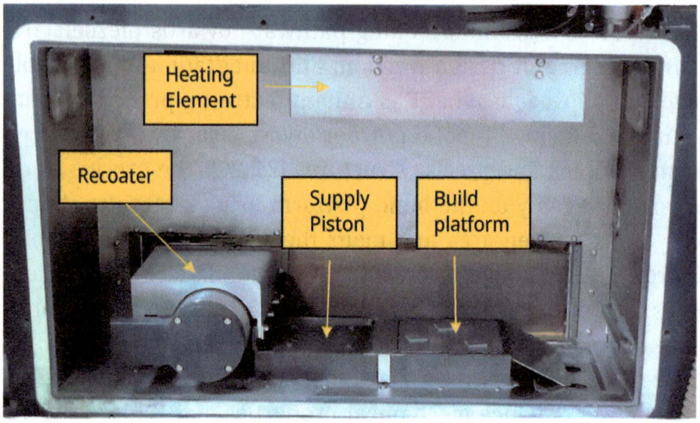

Figure 7.8: LPBF process builds chamber components.

One of the main benefits of LPBF is that it can be used with hybrid and developing multimaterial fabrication approaches. Even though the development of multimaterial processing is still emerging, methods like stop-and-go fabrication, in situ embedding, and graded material transitions have enabled the incorporation of piezoelectric ceramics like (PZT) into metallic matrices. In addition, the parameters of the LPBF process, such as laser power, scan speed, and hatch spacing, provide additional control over the microstructure of the fabricated specimen. It enables tuning of mechanical properties, such as stiffness and damping, thereby improving energy transfer and overall functional performance of the integrated system. In addition to structural integration, the LPBF can also serve as a promising platform for integrating piezoelectric functionality through innovations at the process level. Recent investigations have demonstrated that piezoelectric-based ultrasonic actuation can be added during fabrication to control melt pool behavior, refine grain structures, and minimize defects. This is a change towards more active process control, in which external information is used to modify material properties in real time. Moreover, piezoelectric components may be used for dual purposes, enabling in situ monitoring and nondestructive analysis and allowing detection of defects, residual stresses, and microstructural changes during or after fabrication. This sensing-and-actuation combination aligns well with the concept of hybrid AM. However, issues of material compatibility, thermal incompatibility, and the reliability of the interface remain to be solved. Despite this, the combination of LPBF and piezoelectric materials, hybrid structures, and metastructural design provides a promising direction for the creation of multifunctional, lightweight, and high-performance systems for use in aerospace and biomedical devices, as well as adaptive and energy-harvesting systems.

7.4 Metastructures for enhanced piezoelectric performance

Metastructures use specially designed shapes to achieve better mechanical and functional properties. The geometry of piezoelectric functional structures is decisive in determining the local strain distribution, stress amplification, and energy transfer, which directly affect the efficiency of electromechanical coupling. Along with AM processes like LPBF, these engineered architectures enable the development of highly optimized, lightweight, and multifunctional systems.

7.4.1 Geometry-induced enhancement

The response of piezoelectric materials depends primarily on the strain exerted; hence, any structural design that enhances local strain can be used to enhance their electrical response. Here, metastructured geometries, especially lattice-based ones such as chiral lattices (Figure 7.9), can be used to effectively localize and concentrate strain in certain parts of the component. This allows greater sensitivity and actuation efficiency in sensing without increasing the total volume of material. This method is particularly suitable for components fabricated by LPBF with complex geometries.

Figure 7.9: Schematic of chiral lattices.

7.4.2 Auxetic metamaterials

Auxetic metamaterials belong to a group of materials with a negative Poisson's ratio, which have the special property of expanding laterally when subjected to tensile forces, and contracting uniformly when subjected to compressive forces (Figure 7.10). This nonconventional deformation behavior makes it easier to have greater localization of strains and a more evenly distributed stress, which is why they are better adapted to the incorporation of piezoelectric elements. These architectures have the potential to enhance sensing performance by efficiently transferring strain to embedded piezoelectric elements without compromising structural integrity.

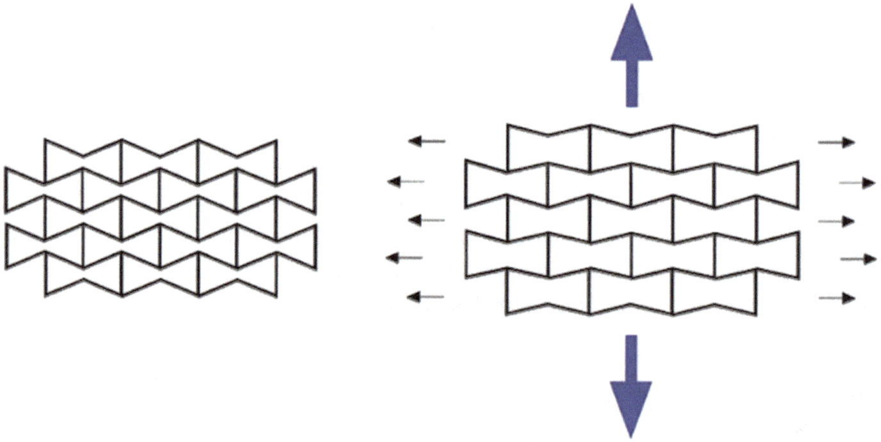

Figure 7.10: Schematic of auxetic lattice structure.

7.4.3 Lattice architectures

Lattice structures can be broadly categorized by their deformation mechanism, which is crucial to piezoelectric performance. Stretch-dominated lattices provide strong rigidity and excellent load sharing and can be used in situations requiring structural stability. Alternatively, lattices that are dominated by bending undergo larger deformations, which can increase the piezoelectric response due to higher strain levels (Figure 7.11). Therefore, lattice topology should be selected based on the required balance between mechanical strength and functional sensitivity.

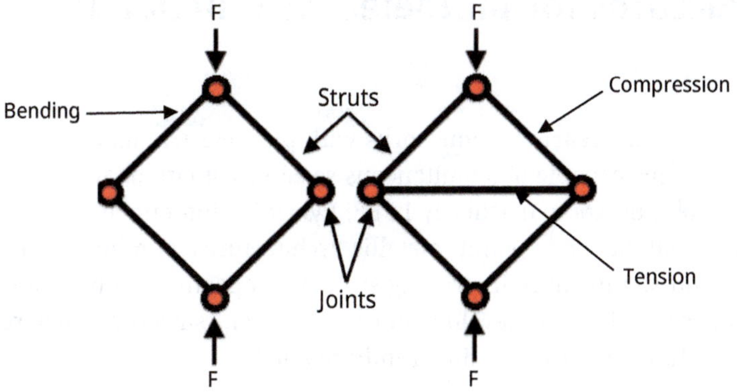

Figure 7.11: Schematic of bending and stretch-dominated structures.

7.4.4 Triply periodic minimal surfaces (TPMS)

An improved form of metastructures employs triply periodic minimal surfaces (TPMS), including the primitive and gyroid geometries, which feature smooth, continuous surfaces and a mean curvature near zero, as illustrated in Figure 7.12. These designs offer various benefits, such as lower stress concentrations, enhanced fatigue resistance, and more even load distribution. TPMS structures offer a promising platform for integrating functional materials into piezoelectric applications because their continuous paths facilitate stress transfer and reduce interface areas. These metastructural design principles provide a robust geometric framework for improving piezoelectric performance, which can be further developed through effective hybrid integration strategies.

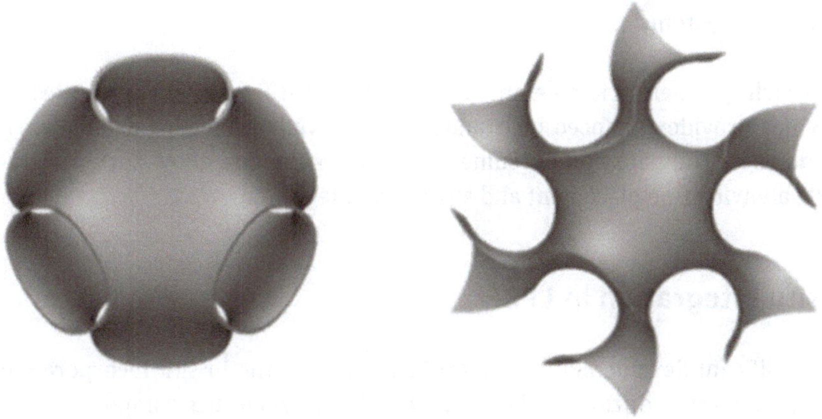

Figure 7.12: Schematic of TPMS lattice unit cells.

7.5 Hybrid structures for piezoelectric functional systems

Hybrid structures combine load-bearing frameworks with piezoelectric materials to create multifunctional systems capable of simultaneous sensing, actuation, and structural support. In the context of AM, particularly LPBF, hybridization enables the direct integration of functional elements within metallic architectures, eliminating the need for postassembly and significantly improving system compactness and performance. This approach is central to the development of intelligent structures, where mechanical and electrical functionalities are inherently coupled.

7.5.1 Integration architectures

Several integration strategies have been developed to incorporate piezoelectric elements into structural systems, each offering distinct advantages depending on the application requirements and manufacturing constraints.

7.5.1.1 Surface-mounted systems

In this technique, piezoelectric elements are attached to the surface of structural components. It is easy to use and common, but its performance is often dictated by the quality of the adhesive interface, which directly controls its strain transfer efficiency and long-term reliability.

7.5.1.2 Embedded systems

In this approach, piezoelectric elements are incorporated into internal cavities or channels, which provide enhanced mechanical contact while protecting against environmental damage. The pause and resume fabrication strategies allow such integration in LPBF, allowing fine placement and structural integration

7.5.2 In situ integration in LPBF

One of the significant developments in hybrid structures is the in situ incorporation of piezoelectric materials during the LPBF process. The piezoelectric components can be temporarily positioned and then embedded into the metallic matrix in subsequent layers by controlling the moment of insertion of such elements. The technique allows for more accurate positioning, better interfacial bonding, and eliminates reliance on external fastening techniques. Moreover, to maintain structural continuity and provide stability during fabrication, design elements such as top caps or arch-shaped supports can be added.

7.5.3 Prestress engineering

Piezoelectric ceramics are brittle and work best under compressive loading. Therefore, the prestress should be introduced under control to ensure stability and a reliable electromechanical response. Prestress in hybrid additively manufactured systems can result from thermal shrinkage during cooling, geometric limitations, or mechanical design features incorporated into the structure. In this regard, LPBF has a distinct advantage, as the thermal gradients generated by the process can be strategically used to produce in situ prestress. Although metastructures improve performance through geometry, hybrid structures enable functional integration at both the material and system levels.

7.6 Piezoelectric smart structures for integrated sensing and actuation

The level of integration of these elements within load-bearing architectures determines the piezoelectric functionality of advanced smart structures. Instead of using external sensors attached to the structure, more recent advances are based on incorporating piezoelectric elements into the structure itself, which provides better strain transfer, better signal quality, and simplifies the entire system. This method is especially effective when coupled with architected geometries, such as lattices and auxetic structures, where deformation can be locally enhanced and directed toward the functional elements housed. Design-wise, embedding may be achieved through internal cavities, layer-wise embedding, or implementation within truss members, enabling the exact location of sensors and actuators within areas where maximum stress or deformation is occurring. These designs allow good electromechanical contact and can be distributed-sensed across the structure without requiring much wiring or external connections.

One of the benefits of embedded piezoelectric systems is that they can monitor structural behavior (stress distribution, vibration response, and damage initiation) in real time and at a local scale. The embedded piezoelectric elements may also be used to create actuators, in addition to sensing, and to provide active control capabilities, including vibration suppression and adaptive stiffness modulation. These capabilities are also improved when combined with metastructured designs. Designed geometries may be designed to focus strain in specific areas, negatively loading the piezoelectric elements by increasing the mechanical contribution. This results in enhanced electrical performance and increased energy conversion efficiency, even at relatively low loading states. The embedded settings are also quite beneficial to structural health monitoring applications by the incorporation of piezoelectric components into architected, additively manufactured structures. This method improves not only sensing and actuation performance but also facilitates the emergence of intelligent structures that can continually monitor and adaptively respond to real operating conditions.

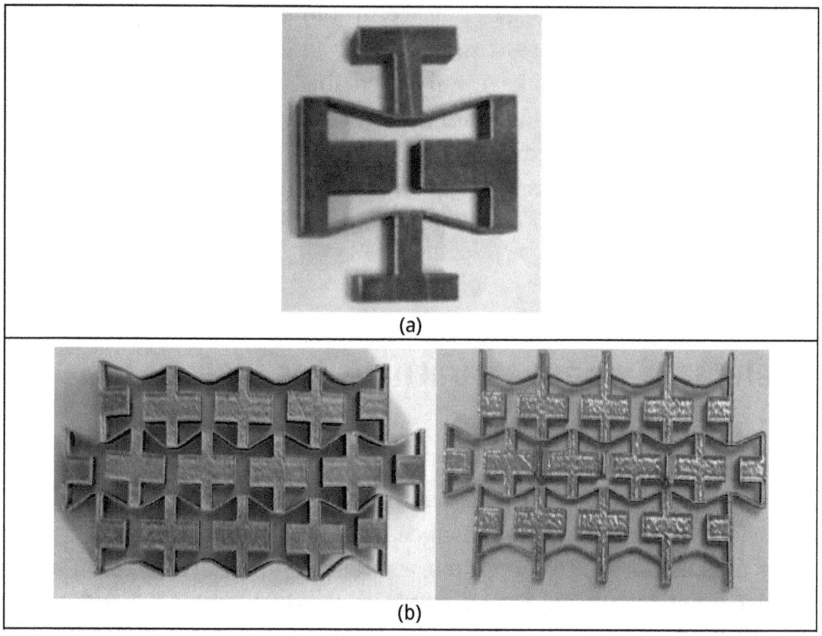

Figure 7.13: (a) T-shaped auxetic component and (b) 3D-printed auxetic walls for piezoelectric smart structure.

Slaby et al. [23] employed an auxetic geometry with a negative Poisson's ratio to allow uniform deformation while maintaining structural integrity, represented by a unit cell of four interlocking T-shaped struts in a honeycomb structure (Figure 7.13a). The final unit cell and auxetic samples were fabricated using SS316L and Ti-6Al-4V via SLM (Figure 7.13b) and used for PZT integration to evaluate voltage response. This study demonstrates a re-entrant auxetic metamaterial that functions both as a load-bearing structure and a self-sensing system.

7.7 In situ integration of piezoelectric components in LPBF

LPBF has become one of the prominent metal AM techniques, capable of creating very complex, lightweight parts by melting metal powders layer by layer. In addition to geometric flexibility, its layer-wise fabrication approach offers a unique opportunity to integrate functional elements into structural components during the building process. This possibility can expand the role of LPBF as a manufacturing method into a platform for multifunctional systems development. The ability to pause the build and restart it is one of the most important benefits of LPBF in this regard, as it enables a temporary pause to add functional components and then resume fabrication. This allows embedding piezoelectric devices, including stacks, into metallic frameworks with high positioning specifications. Besides that, the thermal cycles of LPBF can also be used to their advantage; during solidification and cooling, the material can undergo thermal contraction, providing a useful prestress to the embedded piezoelectric components, which is vital to their stability, durability, and efficient electromechanical response. This integration is especially beneficial for applications in metallic structures that require vibration control. The direct conversion of mechanical vibrations into electrical signals is facilitated by piezoelectric elements, which can be used for active or passive damping with external electronics. It is a very appropriate method for vibration mitigation in lightweight, load-bearing systems without loss of structural stiffness. Although these are the benefits, incorporating piezoelectric components into metal-based AM is not easy.

In contrast to polymer-based systems, in which embedding is comparatively easy, metallic environments have high processing temperatures and require demanding structural requirements. To address these limitations, recent studies have focused on in situ integration strategies in the LPBF process. These methods, including placing piezoelectric elements during fabrication and then embedding them with the other layers that follow, have shown promising results. Through design adaptations, including the addition of top caps or enclosure capabilities, controlled prestress can be developed due to thermal shrinkage, making the use of extra mechanical parts unnecessary. Simultaneously monitoring parameters such as force and voltage can also be very helpful in understanding the development of prestress and the response of embedded elements.

Weber et al. [24] demonstrated that LPBF can incorporate piezoelectric stacks directly into the complex geometry using its layer-wise access, and that the component can be placed during fabrication, as shown in Figure 7.14. The research highlights the application of thermal shrinkage of the structure to induce internal prestress, without the use of an external mechanism.

Figure 7.14: (a–b) Piezoelectric stack integration in AM truss structures.

7.8 Ultrasonic piezoelectric actuation for microstructure control in LPBF

Piezoelectric actuation in LPBF provides an active means to control microstructural development during fabrication. One common issue in LPBF is the formation of anisotropic microstructures, mostly owing to steep thermal gradients and high solidification rates, which favor columnar grain propagation in the build direction. This anisotropy can impair the mechanical reliability of demanding applications. Ultrasonic actuation is possible, and piezoelectric transducers allow active control of melt-pool dynamics and real-time control of solidification. Under this method, transducers are added to the build platform or substrate to generate high-frequency vibrations. These ultrasonic waves are reflected in the melt pool, producing localized cavitation effects and acoustic streaming. This leads to an increase in fluid flow in the melt pool, a disruption of the solidification front, and nucleation, which contributes to a shift toward finer, more equiaxed grain structures. The induced vibrational energy can alter the environment of solidification, even with relatively small displacement amplitudes. The success of ultrasonic-assisted LPBF is determined by the efficient transmission of vibrational energy across interfaces, especially at the transducer, substrate, and powder layers. In addition to grain refinement, ultrasonic actuation helps achieve melt-pool stability, reduce defects, and increase material uniformity, resulting in greater mechanical isotropy and reduced cracking potential.

Griffin et al. [27] created a piezoelectric transducer in the form of a fixture that efficiently transmits ultrasonic waves through the LPBF build plate, enabling active control over solidification behavior, as shown in Figure 7.15. Their findings demonstrated that a transducer design was optimized to enable near-transmission of energy, which greatly influenced melt-pool morphology and reduced defects. This method

Figure 7.15: (a) Ultrasonic piezoelectric actuation via transducer and (b) transducer mounted on aluminum build plate.

demonstrates the potential of ultrasonic actuation to enhance control of microstructure and in situ monitoring of the LPBF process.

7.9 Metastructures and multimaterial architectures for piezoelectric functional systems

LPBF is opening new possibilities by enabling metastructures and multimaterial architectures to design piezoelectric functional systems with improved performance and tunability. In their earlier publication, Kalia et al. [7, 8] showed that the LPBF process can be used to produce a compositionally graded structure, with Ti-6Al-4V at the top and a lattice interlayer of Ni-625 and 17-4 PH stainless steel at the bottom. Based on this research, a pilot experiment was performed on gas-atomized 17-4 PH stainless steel, Ni-625, and Ti-6Al-4V powders (15–53 μm) to fabricate multimaterial auxetic lattice structures using an LPBF system via a controlled powder-swapping method, as shown in Figure 7.16(a–b). This process enabled the sequential deposition of various materials in a single build with good geometric control. The study also comprises various lattice-based metastructures like auxetic, Weaire-Phelan, and octet-truss that, besides reducing weight, contribute to the overall functionality of the structures. Strain can be directed and confined to specific parts of the structure by selectively shaping the inner lattice. This will be especially useful in piezoelectric integration, since locally amplified strain will directly enhance the electrical response, enabling higher output under a comparatively low external load. Along with geometric control, multiple metallic materials within a single structure broaden design choices. By combining different alloys (Ti-6Al-4V, Ni-625, 17 4 PH SS), it facilitates tuning of mechanical properties, including stiffness, strength, and ductility.

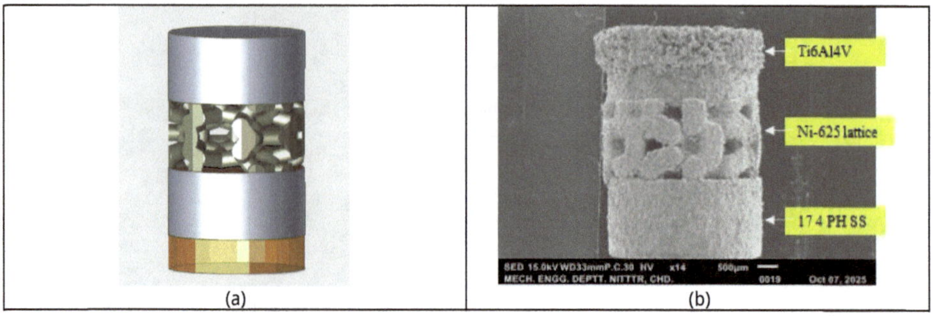

Figure 7.16: (a) Schematic for Weaire-Phelan lattice integrated structure and (b) multimaterial 3D-printed lattice integrated structure.

There is scope to further enhance piezoelectric performance by integrating functional ceramic phases directly into the material system. A potential solution includes developing a composite feedstock reinforced with stainless steel (SS316) and barium titanate ($BaTiO_3$). A starting point for a composition of 95 wt% stainless steel and 5 wt% $BaTiO_3$ can be an initial production of a composition that balances structural integrity and functional response. Issues such as thermal mismatch, particle dispersion, and phase stability during LPBF processing are concerns that need to be addressed to produce intrinsically piezoelectric metastructures with single-step fabrication. Here, multimaterial processing, architected geometries, and functional material integration offer a promising direction to advanced piezoelectric systems. Such structures maintain mechanical efficiency and provide integrated sensing and energy-conversion capabilities, making them well-suited for advanced smart engineering applications.

7.10 Conclusions

1. The integration of metastructures and multimaterial architectures through AM has significantly expanded the range of design options for piezoelectric functional systems. By combining material selection with geometry-driven design, it is now possible to achieve enhanced strain localization and improved electromechanical response within structurally efficient components.
2. The hybrid integration strategies, including both in situ embedding and ex situ placement of piezoelectric elements, offer flexible ways to integrate sensing and actuation capabilities. These approaches enable the development of multifunctional structures capable of performing both mechanical and functional roles.
3. The advanced approaches, such as ultrasonic piezoelectric actuation in LPBF, demonstrate strong potential for active control over melt pool behavior and microstructure

evolution. This improves material uniformity and reduces defects, while also opening new opportunities for process-level functionalization in AM.

4. Despite challenges related to material compatibility, interface stability, and process optimization, recent progress indicates a shift toward intelligent and adaptive systems. The integration of structure, sensing, and actuation is expected to play a key role in future applications such as vibration control, structural health monitoring, and smart engineering systems.

References

[1] Bertoldi K, Vitelli V, Christensen J, Van Hecke M. Flexible mechanical metamaterials. Nature Reviews Materials. 2017 Oct 17;2(11):1–1.

[2] Kadic M, Milton GW, Van Hecke M, Wegener M. 3D metamaterials. Nature Reviews Physics. 2019 Mar;1(3):198–210.

[3] Burresi M, Diessel D, Oosten DV, Linden S, Wegener M, Kuipers L. Negative-index metamaterials: Looking into the unit cell. Nano letters. 2010 Jul 14;10(7):2480–83.

[4] Valipour A, Kargozarfard MH, Rakhshi M, Yaghootian A, Sedighi HM. Metamaterials and their applications: an overview. Proceedings of the Institution of Mechanical Engineers, Part L: Journal of Materials: Design and Applications. 2022 Nov;236(11):2171–210.

[5] Cui J, Zhang L, Gain AK. A novel auxetic unit cell for 3D metamaterials of designated negative Poisson's ratio. International Journal of Mechanical Sciences. 2023 Dec 15;260:108614.

[6] Noronha J, Leary M, Brandt M, Qian M. AlSi10Mg hollow-strut lattice metamaterials by laser powder bed fusion. Materials Advances. 2024;5(9):3751–70.

[7] Kalia B, Singh R. On 3D Printing of 17-4 PH SS Rectangular Waveguides with Laser Powder Bed Fusion. National Academy of Sciences Letters. 2025a. https://doi.org/10.1007/s40009-025-01642-1.

[8] Kalia B, Singh R, Pabla BS. On Interfacial Characterization and Thermodynamic Analysis of Ni-625-17-4 PH Stainless Steel-Based Functionally Graded Material Fabricated by Laser Powder Bed Fusion. Journal of The Institution of Engineers (India): Series C. 2025c;106:1165–71.

[9] Kalia B, Singh R, Pabla BS. Investigating strategies for the fabrication of composition gradient-based functionally graded materials processed by powder bed fusion. Arabian journal for science and engineering. 2025b. https://doi.org/10.1007/s13369-025-10898-0.

[10] Fan Y, Wang Y. The effect of negative Poisson's ratio on the low-velocity impact response of an auxetic nanocomposite laminate beam. International Journal of Mechanics & Materials in Design. 2021 Mar 1;17(1).

[11] Kalia B, Singh R, Pabla BS. On the Effect of Unit Cell Dimensions in Octet Auxetic Structure Fabricated by Laser Powder Bed Fusion. Journal of Materials Engineering and Performance. 2026;35:5444–56.

[12] Papadopoulou A, Laucks J, Tibbits S. Auxetic materials in design and architecture. Nature Reviews Materials. 2017 Dec 5;2(12):1–3.

[13] Lira C, Innocenti P, Scarpa F. Transverse elastic shear of auxetic multi-re-entrant honeycombs. Composite Structures. 2009 Oct 1;90(3):314–22.

[14] Plewa J, Płońska M, Lis P. Investigation of modified auxetic structures from rigid rotating squares. Materials. 2022 Apr 13;15(8):2848.

[15] Zhang Z, Gu Y, Wu H, Chen Q. Investigation on the energy absorption characteristics of novel graded auxetic re-entrant honeycombs. Composite Structures. 2025 Jan 15;352:118633.

[16] Srivastava C, Bhola L, Mahesh V, Guruprasad PJ, Petrinic N, Scarpa F, Harursampath D, Ponnusami SA. Exploiting nonlinearities through geometric engineering to enhance the auxetic behaviour in re-entrant honeycomb metamaterials. Scientific Reports. 2023 Nov 27;13(1):20915.

[17] Teng XC, Ren X, Zhang Y, Jiang W, Pan Y, Zhang XG, Zhang XY, Xie YM. A simple 3D re-entrant auxetic metamaterial with enhanced energy absorption. International Journal of Mechanical Sciences. 2022 Sep 1;229:107524.

[18] Dudek KK, Gatt R, Mizzi L, Dudek MR, Attard D, Evans KE, Grima JN. On the dynamics and control of mechanical properties of hierarchical rotating rigid unit auxetics. Scientific reports. 2017 Apr 26;7 (1):46529.

[19] Najafi M, Ahmadi H, Liaghat G. Experimental investigation on energy absorption of auxetic structures. Materials today: proceedings. 2021 Jan 1;34:350–55.

[20] Kumar V, Singh R, Ahuja IS et al. 3D Printed Meta-structure-Inspired Sensors of PVDF–Graphene–Mn-Doped ZnO for Heritage Structures. Natl. Academy of Science Arts and Letters, 2023, 46, 199–202.

[21] Wei T, Lu F, Zhang C, Huang Y, Rui X, Zhu Y. Energy absorption of 3D assembled auxetic meta-structure with compression-twisting effect. In Structures 2025 Mar 1 (Vol. 73, p. 108482). Elsevier.

[22] Santosa SP, Maulana R, Setiawan AR. Energy absorption performance of additively manufactured 3D elliptical anti-chiral auxetic metastructures with 20MnCr5 steel. Next Materials. 2026 Apr 1;11:101918.

[23] Slaby V, Bajer J, Marcian P, Hrstka M, Hadas Z. Novel metamaterial platform with piezoelectric sensors for self-sensing mechanical support. International Journal of Mechanical Sciences. 2026 Feb;11:111387.

[24] Weber R, Seydel S, Spierings A, Wegener K. In-situ integration of piezoelectric stacks with laser-based powder bed fusion. Procedia CIRP. 2022 Jan 1;113:329–34.

[25] Singh R, Singh R, Ahuja IPS. Comprehensive Investigation of Scan Strategies and Infill Patterns on the Flexural, Thermal, and Microstructural Behavior of Powder Bed Fusion-Laser Beam-Processed 17-4 PH Stainless Steel. Journal of Materials Engineering and Performance. 2025a. https://doi.org/10.1007/s11665-025-12871-2.

[26] Singh R, Singh R, Ahuja IPS. On thermomechanical properties of 17-4 PH stainless steel processed by using laser powder bed fusion: Effect of scanning strategies and infill patterns. Journal of Materials Engineering and Performance. 2025b. https://doi.org/10.1007/s11665-025-12662-9.

[27] Griffin C, Mei H, Karna S, Zhang T, Giurgiutiu V, Yuan L. Exploring Piezoelectric Actuation towards Its Applications in Laser Powder Bed Fusion Additive Manufacturing. Sensors. 2024 Jun 7;24(12):3704.

Abhishek Barwar, Prateek Kala, Rupinder Singh*

Chapter 8
Investigations of the sensing and morphological properties of 3D-printed piezoelectric functional structures

Abstract: Recently, piezoelectric polymers and piezocomposites have been utilised widely for energy harvesting or sensing applications due to their capability to convert the mechanical strain into electrical energy. Several studies have reported the effect of nanostructures on the dielectric properties of thin piezoelectric films, but hitherto little has been reported on the effect of 3D-printed gradient-based functional structures on the piezoproperties of implantable sensors. In this chapter, a comparative analysis has been performed between two types of functionally graded materials (FGMs), i.e., a metastructure-based and a gradient-based functional structure for the sensing application. The sensor targeted in this chapter operates on the principle of a microstrip patch antenna and comprises three layers: a patch (conductive), a substrate (dielectric), and a ground plane (conductive). The intermediate layer, or substrate, of the sensor has been further divided into three layers to assess the effect of FGMs on the sensor's sensing characteristics. Initially, both functional structures were designed in Ansys high-frequency structure simulator (HFSS), and the sensor's radio-frequency characteristics were recorded. To observe the effect of gradient-based FGM, barium titanate ($BaTiO_3$) has been added to the polyvinylidene fluoride (PVDF)-based matrix, i.e., PVDF reinforced with hydroxyapatite (HAP) and chitosan (CS), whereas simply introducing the metastructure in the intermediate layer of the substrate constitutes the other FGM. Since the sensor has been designed for the industrial, scientific, and medical (ISM) band (2–2.5 GHz), the sensor with metastructure-based FGM shows more promising results comparatively. Further, morphological analysis of the 3D-printed samples has been performed for both FGMs. Based on the comparative analysis, the effective one will be used for piezoelectric sensor applications.

Keywords: Functionally graded material, implantable sensor, radio-frequency, ISM band, piezoelectric structures

***Corresponding author: Rupinder Singh,** Department of Mechanical Engineering, National Institute of Technical Teachers Training and Research, Chandigarh 160019, Chandigarh, India, e-mail: rupindersingh@nitttrchd.ac.in
Abhishek Barwar, Prateek Kala, Department of Mechanical Engineering, BITS Pilani, Pilani Campus, Rajasthan, India

 | https://doi.org/10.1515/9783112218846-008

8.1 Introduction

In recent years, with the increase in demand for materials with tailored functionalities across fields such as electronics, aerospace, biomedical, and applied sciences, polymeric piezoelectric materials, due to their distinctive characteristics and numerous advantages, have appeared as an effective alternative to the pre-existing ceramic-based materials [1]. Key attributes such as low density, lightweight, conformability, ease of processing, and mechanical pliability make them suitable for processes such as solution casting or extrusion [2, 3]. Certain piezoelectric polymers exhibit high biocompatibility, thereby expanding their utility for biomedical applications such as sensor fabrication and the development of other medical devices. In addition to these intrinsic properties, these material properties can be tailored by adding inorganic fillers or blending with certain polymers to fabricate a piezocomposite, thereby further optimising their performance for certain applications [4, 5]. Although the function of a piezoelectric material is to convert the applied mechanical energy into electrical output, but the application areas, such as the curved complex structures, won't be suitable for the piezoelectric ceramics because of their certain limitations, like brittleness, stiffness, high thickness or weight; therefore, piezoelectric polymers have found a lot of applications nowadays [6, 7]. Recently, functionally graded materials (FGMs) have attracted significant interest due to their novel design features for improving the performance of piezoelectric materials; they are also easily adaptable to complex mechanical environments [8]. FGMs exhibit changes in their microstructure and composition, thereby allowing their properties to be altered over time [9, 10]. These are used in high-strength, high-temperature applications, such as medical implants and the aerospace industry, where their unique physical properties help reduce stress concentrations and manage heat gradients [11, 12]. Inclusion of FGMs into piezoelectric materials enables new applications such as energy harvesting and adaptive structures, as these materials provide controlled electromechanical coupling, which is desirable for sensors or actuators [13, 14]. The processing of piezoelectric materials via 3D printing is still in its early stages, as traditional processing techniques such as injection moulding, cast moulding, and dry pressing pose challenges for fabricating complex, intricate, or high-precision geometries [15]. 3D printing can overcome these challenges by enabling greater design flexibility and reducing material waste [16]. The 3D-printed structure offers improved electrical performance, better strain control, enhanced piezoelectric coefficient, design flexibility, and greater sensitivity to piezoelectric materials, which is difficult to achieve with conventional methods [17]. Fabricating the complex structures of FGMs is less challenging when the piezoelectric material is processed via additive manufacturing, while retaining its physical properties. Polyvinylidene fluoride (PVDF) is one of the piezopolymers that is semicrystalline in nature, mostly made up of fluorine and carbon [18]. The strong electronegativity of the fluorine atoms in the chain induces a significant dipole moment, which thereby improves the piezoelectric effect [19]. It has several applications due to its excellent stability, thermal resistance, high flexibility, biocompatibility, and high piezoelectric coefficient [20]. Also, it has certain limitations, such as its hydrophobic nature, which limits cell adhesion and proliferation in implantable applications, a lack of antimicrobial properties, and relatively lower biodegradability [21]. Reinforcing inorganic materials such as hydroxyapatite (HAP) and chitosan (CS) can enhance the physical properties of PVDF, including bioactivity, antibacterial properties, mechanical strength, improved cell proliferation, and biocompatibility [22, 23]. Also, some studies reported barium titanate ($BaTiO_3$) reinforcement to enhance the piezoelectric properties of the polymer matrix. Piezocomposites connect the flexible polymer to the piezoelectric ceramic in a specific mixing ratio, with distinct spatial distributions when processed using different techniques [24]. Recently, 3D printing emerged as an alternative approach for producing highly oriented fibres with a higher degree of crystallinity and piezoelectric chain morphology [25]. PVDF has several applications, including energy harvesting, nanogenerators, and lab-on-a-chip sensors. MPA, due to its significant advantages such as ease of fabrication, design flexibility, low cost, and high-quality communication, is suitable for the fabrication of implantable sensors [26]. PVDF, due to its significant properties and piezoelectric nature, is suitable for the fabrication of implantable MPA-based sensors [27]. Recently, piezoelectric polymers and piezocomposites have been widely utilised for energy harvesting and sensing applications due to their ability to convert mechanical strain into electrical energy. Several studies have reported the effect of nanostructures on the dielectric properties of thin piezoelectric films, but hitherto little has been reported on the effect of 3D-printed gradient-based functional structures on the piezo-properties of implantable sensors.

8.2 Bibliometric analysis

Going forward, to further explore the existing literature and better define the research gap, a bibliometric analysis has been performed on data extracted from the Scopus database over the past two decades. The keywords searched in the Scopus database are piezoelectric materials, FGMs, and 3D printing. As a result, several documents relevant to these terms appeared and can be further utilised to create certain bibliometric charts. Figure 8.1(a) highlights the number of documents published over the years in the field of 3D printing of piezoelectric functional structures. The data show that initially, the graph rises slowly up to 2008, but after 2009, a larger number of documents are reported, most likely due to the expiration of the Stratasys patent for fused deposition modelling, which eventually led to the rise of open-source 3D printing companies. Therefore, more materials are being explored for 3D printing a range of parts for multiple applications. Over the last 5 years, an average of 135 documents per year have been published on 3D printing of piezoelectric structures. Figure 8.1(b) shows the distribution of documents across different subject areas: 40% belong to the engineering domain, followed by materials science (23.2%) and physics and applied sciences (15.8%). This, therefore, highlighted the demand for 3D printing to develop piezoelectric nanostructures for fabricating flexible membranes. Moving ahead, Figure 8.1(c) presents the nature of the documents reported over the past two decades, along with their % share in a pie chart. It was observed that the most reported documents are of the article type (82.4%), highlighting the globalisation of 3D-printed piezoelectric structures. Figure 8.1(d) highlights the contribution of the top ten countries in the field of 3D printing of piezoelectric structures, with China leading the list by publishing over 700 documents in the past two decades, followed by Iran (>400), and India (>200). Further, Figure 8.1(e) shows the top journals contributing to the field since 2001, and it was found that "Mechanics of advanced materials and structures" accounted for the highest number of documents, i.e. 14 in 2025. This data provides information on recent trends in piezoelectric structures fabricated via 3D printing and has proven beneficial for this research.

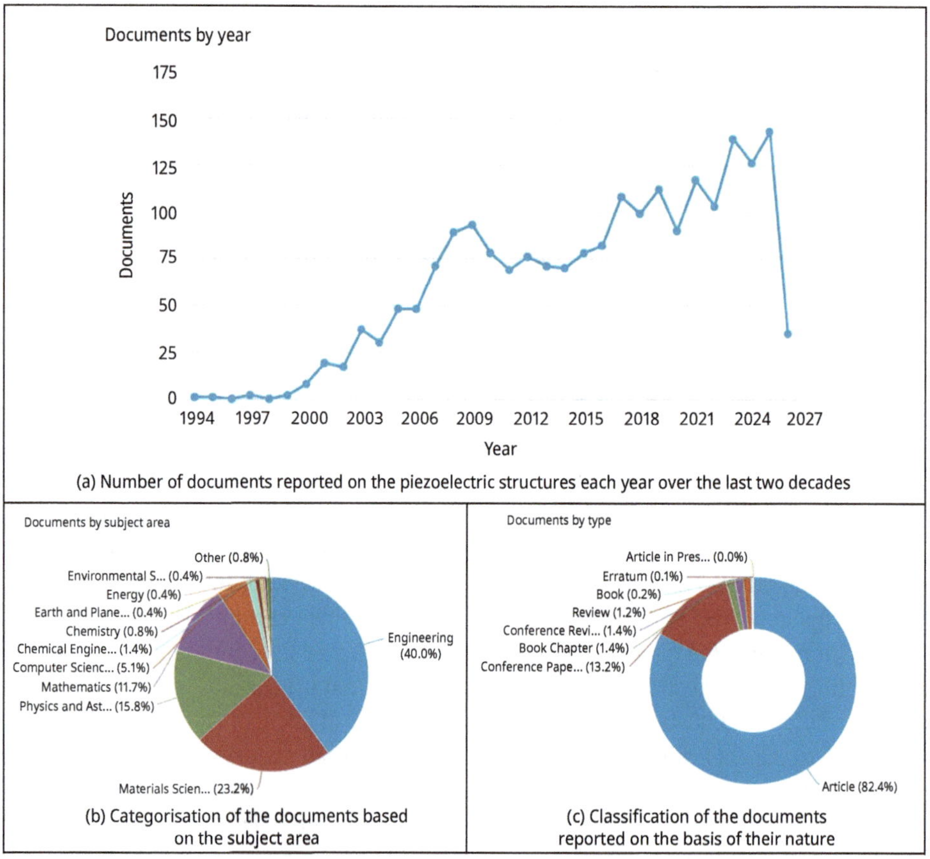

(a) Number of documents reported on the piezoelectric structures each year over the last two decades

(b) Categorisation of the documents based on the subject area

(c) Classification of the documents reported on the basis of their nature

Figure 8.1: Bibliometric analysis of the piezoelectric structures processed through 3D printing based on the data extracted from the Scopus database.

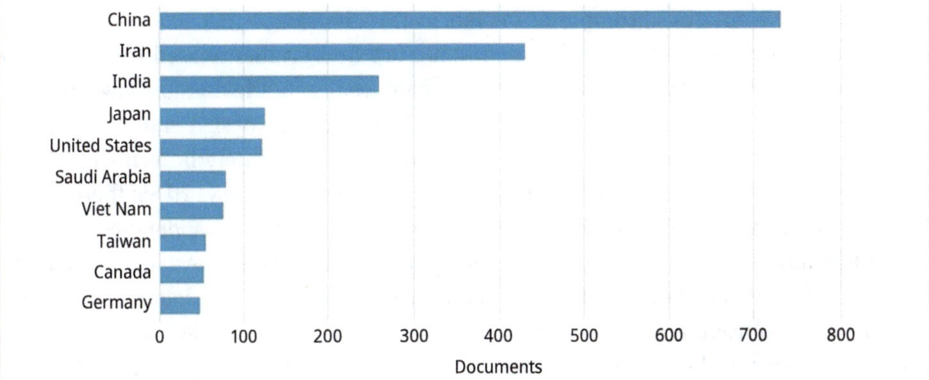

(d) Top contributing countries in the 3D printing of functional piezostructures

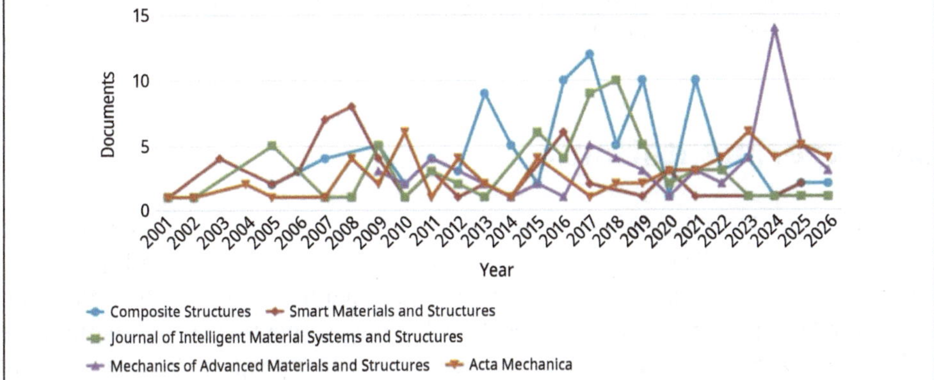

(e) Top contributing journals in the field over the last two decades

Figure 8.1 (continued)

8.3 Methodology

To delve deeper into the piezoelectric functional structures based on fabrication technique, sensing characteristics, and morphological features, this chapter explores two different types of FGMs (Figure 8.2). Primarily, the piezoelectric material composition has been selected to target the fabrication of an implantable sensor for remote patient health monitoring. PVDF possesses excellent chemical and thermal resistance, as well as mechanical robustness and biocompatibility, in addition to its piezoelectric nature. On the contrary, PVDF also presents some challenges, such as its hydrophobic nature, which limits proliferation and cell adhesion for implantable applications. HAP and CS are commonly used materials for scaffolding or implant applications due to their enhanced bioactivity and cell proliferation at the host site, as well as their porous structures that promote osseointegration. The HAP and CS were reinforced at 8% and 2%, respectively, in PVDF to develop a piezocomposite for fabricating the sensor [27]. After selecting the composition, two types of FGMs, i.e., metastructure and gradient-based (the substrate was divided into three layers, and the material composition of the topmost layer is inclusion of 2% $BaTiO_3$ in 98% of PVDF-HAP-CS), were chosen for the design and fabrication of the dielectric layer of the sensor and were characterised separately in terms of RF and morphological parameters. Initially, the sensor was designed in Ansys HFSS; therefore, to highlight the inclusion of both types of FGMs in two different cases, a detailed comparison of the output parameters obtained from sensor simulations has been conducted, and the effect of FGMs has been analysed. Based on the simulations, the final sensor dimensions have been used to fabricate the sensor substrate via the material extrusion (MEX) process, optimising the process parameters. Subsequently, the morphology of the prepared samples through both FGMs has been examined. Following that, a comparison based on sensing and morphological investigations was conducted to better select the FGM for sensor fabrication or other piezoelectric applications.

8.4 Experimentation

To prepare the PVDF-based piezocomposite, the HAP, CS, and $BaTiO_3$ nanoparticles are mixed with PVDF granules in a ball-milling setup at 80 rpm for 3–4 h [22]. Once uniform mixing was achieved, the compound was used to prepare the feedstock filaments for MEX printing of the samples. Initially, the compound was poured into the twin-screw extruder (TSE) for uniform blending of the nanoparticles in the PVDF-based matrix, and the operating parameters of TSE are screw temperature 210 °C, and the screw speed was 100 rpm. Following that, to address the non-uniformities in the diameter, the prepared filament was shredded into small lengths (average of 5 mm) and further processed through the single-screw extruder (SSE) at the operating parameters such as screw temperature 210 °C and screw speed 6 rpm, and the output diameter was 1.70 ± 0.05 mm. A separate filament was prepared with the composition PVDF-HAP-CS (98%) and $BaTiO_3$ (2%) for fabricating the topmost layer of the substrate for the gradient-based FGM, whereas the other layers will be processed with the PVDF-HAP-CS filaments. Once the uniform filament diameter was achieved, the filaments were further processed through the MEX process to fabricate substrates for different FGMs.

The samples have been 3D-printed on the Prusa i3 MK3S + MEX printer. The operating parameters of the MEX printer, such as printing speed, flow rate, bed temperature, and layer height, were optimised to achieve continuous 3D printing with minimal errors, such as bed adhesion (a commonly reported issue during processing PVDF) and discontinuous piezocomposite flow. The metastructure-based FGM is 3D-printed in a single run, with the intermediate substrate layer using a rectangular-grid

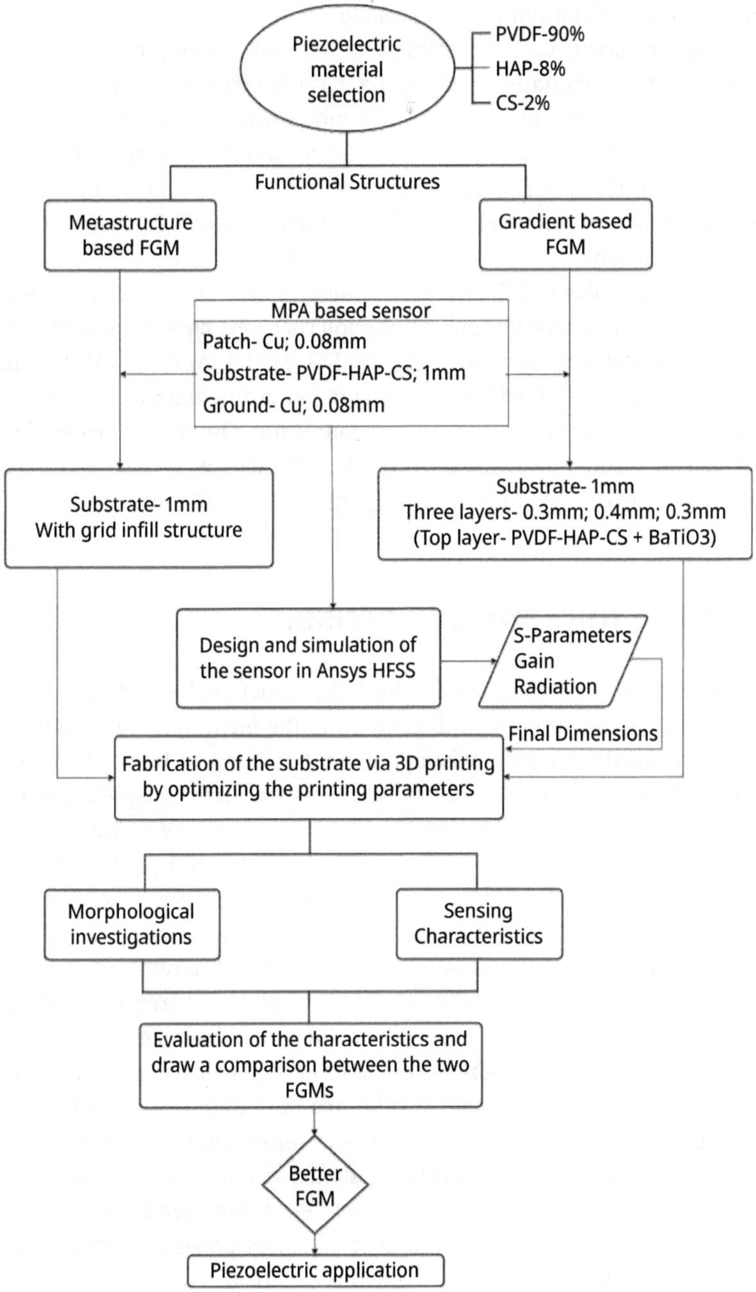

Figure 8.2: Steps involved in this study.

infill pattern, while the top and bottom layers are solid. On the contrary, for the gradient-based FGM, the bottom layer, with a thickness of 0.6 mm, is composed of PVDF-HAP-CS, whereas the top layer contains traces of $BaTiO_3$ (2%) from the earlier developed piezocomposite. The layer height was set to 0.16 mm; printing speed, 50%; nozzle end temperature, 230 °C; heating bed temperature, 90 °C; nozzle diameter, 0.6 mm; and the flow rate was set to 100% during 3D printing of the samples. Once 3D printing was complete, the samples were removed from the magnetic bed and further utilised for morphological investigations.

To understand the morphology of the prepared samples and the effect of fabrication on the sensing characteristics of the sensors, the micrographs have been captured at different magnification scales using FESEM (Make: FEI, Model: Apreo Lo Vac) with an EDS system (AZtecOne, Make: Oxford Instruments) to better understand the composition of the various elements present in the piezocomposite. Other key properties, such as the average surface roughness and a 3D-rendered image, were also extracted from the FESEM images to further examine the morphology of the 3D-printed samples.

8.5 Design of the implantable sensor

The sensor shown in this chapter operates on the MPA principle, i.e. a device commonly used for transmitting and receiving information in the form of electromagnetic (EM) waves in a design-specific frequency band. Since it is going to be used for the implantable application, the industrial, scientific, and medical (ISM) frequency band (2–2.5 GHz) is desirable for designing this sensor. The basic structure of this sensor comprises three units, i.e., a metallic patch, a substrate, and a ground plane. And the sensor, upon receiving the input electrical signal, emits radiation in the form of EM waves from the edges of the metallic patch over a specific frequency range. For designing this sensor, its dimensions have been evaluated based on input parameters such as a substrate height of 1 mm, an input impedance of 50 Ω, a resonant frequency (f_r) of 2.45 GHz, and a dielectric constant (ε_r) of 1.97. To transport the input signal, a microstrip line feed has been provided to this sensor due to its ease of construction and minimal signal loss. Well-established mathematical relations (8.1)–(8.4) [28] were used to calculate the dimensions of the patch and the feed. The sensor patch has a width of 50.2414 mm and a length of 43.0874 mm, and the assigned material was copper (Cu), with a finite conductivity for the boundary conditions. Feed line's length and width are 13.503 and 3.5 mm, respectively, and placed over an offset position to match the input impedance of 50 Ω. The ground plane and the substrate were measured as 70 × 70 mm and, similar to the patch ground plane, were assigned a finite-conductivity boundary condition with Cu as the conductive material. The substrate thickness (1 mm) is divided into three layers, i.e., 0.3, 0.4, and 0.3 mm, respectively, for both the FGMs, and assigned with the material PVDF-HAP-CS to the bottom two layers in case of gradient-based FGM, whereas the topmost layer was assigned with the PVDF-HAP-CS + $BaTiO_3$, whereas PVDF-HAP-CS for all the three layers in case of meta-structure based FGM has been taken (Figure 8.3(a–e)):

$$\varepsilon_{\text{eff}} = \left[\frac{(\varepsilon_r + 1)}{2}\right] + \left[\left(\frac{(\varepsilon_r - 1)}{2}\right)\left(1 + 12\frac{h}{w}\right)^{-0.5}\right] \tag{8.1}$$

$$\Delta L = 0.42h\left[\frac{(\varepsilon_{\text{eff}} + 0.3)\left(\frac{w}{h} + 0.264\right)}{(\varepsilon_{\text{eff}} - 0.258)\left(\frac{w}{h} + 0.8\right)}\right] \tag{8.2}$$

$$L = \frac{1}{2f_r\sqrt{(\varepsilon_{\text{eff}}\varepsilon_o\mu_o)}} - 2\Delta L \tag{8.3}$$

$$W = \frac{C}{2f_r}\sqrt{2/(\varepsilon_r + 1)} \tag{8.4}$$

where ε_{eff} is the effective dielectric constant, ε_r is the dielectric constant, w is the width of the transmission line, h is the substrate height, ΔL is the fringe factor, W and L are the width and length of the patch, respectively, f_r is the resonant frequency, and C is the velocity of light.

8.6 Results and discussion

8.6.1 R-F characteristics of the sensor

The simulated behaviour of the sensor has been further analysed by plotting various R-F characteristics in the ISM frequency band, namely the scattering parameter (S_{11}), gain, and radiation efficiency (η_{Rad}) against frequency. Looking at each, the S_{11} versus f_r results of metastructure-based FGM are highlighted in Figure 8.4(a), showcasing an impressive value of S_{11} (−25.52 dB) corresponding to f_r of 2.37 GHz, whereas, looking at the S_{11} versus f_r results of gradient-based FGM, it was observed that the sensor resonated at 2.02 GHz while having a S_{11} of −17.93 dB (Figure 8.4e). The metastructure-based substrate possesses a low value of the ε_r due to the infill structure, therefore resonated at a higher frequency, whereas in the case of gradient-based FGM, due to the inclusion of the $BaTiO_3$, the overall ε_r of the substrate has been increased [29], causing it to resonate at a lower value (2.02 GHz) comparatively. The maximum sensor gain values in both cases are 7.27 and 6.35 dB, respectively (Figure 8.4b and f). This highlights the sensors' ability to communicate over a greater distance. While plotting the gain against frequency, sharp peaks were observed at 2.25 GHz in both cases, with corresponding gain values of 6.45 and 6.23 dB, respectively (Figure 8.4c and g). Also, looking at η_{Rad}, it was observed that the sensor with metastructure-based FGM is more efficient (0.72) than the other FGM (0.70) (Figure 8.4d and h). More losses were

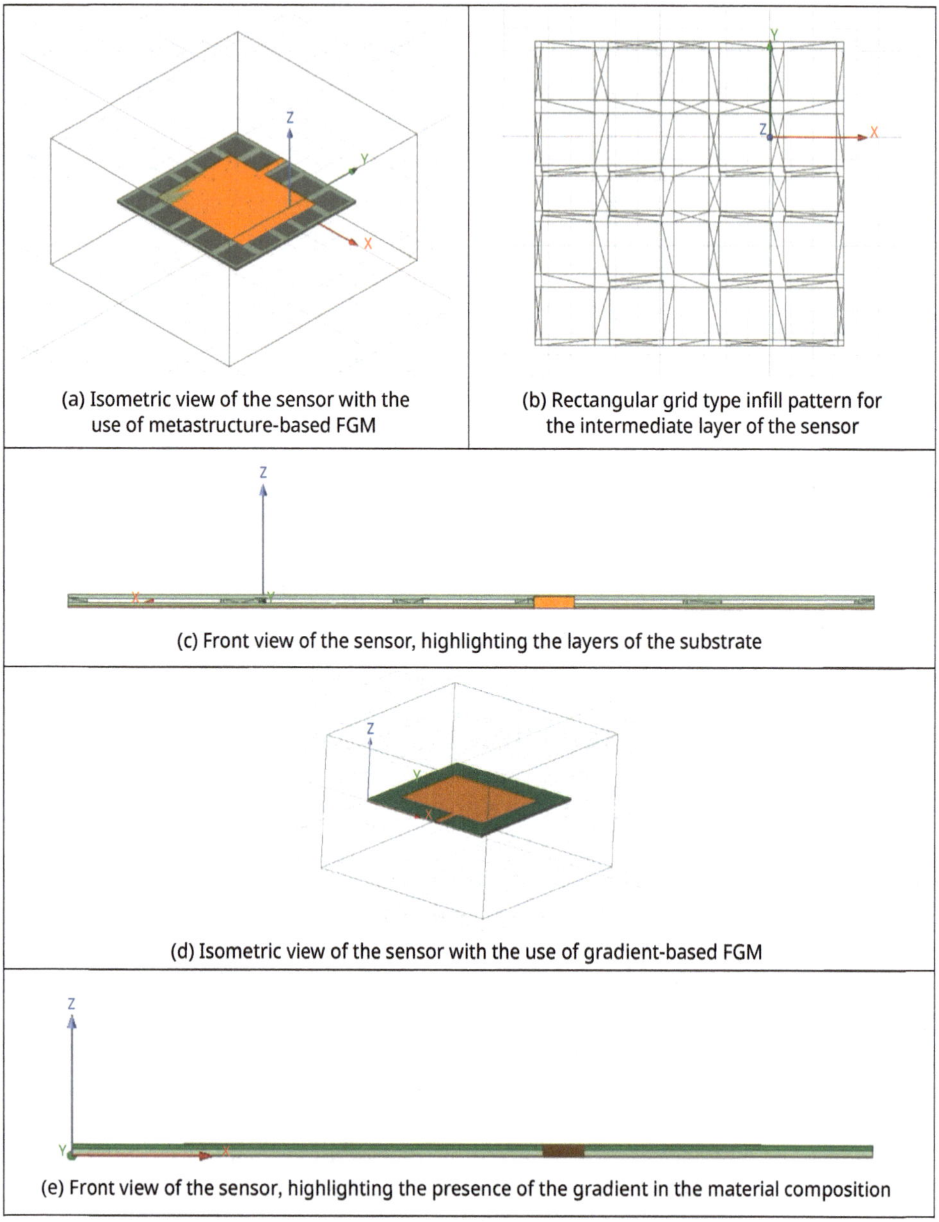

Figure 8.3: Design of the implantable sensor utilising two types of FGMs.

observed with the inclusion of $BaTiO_3$ in the piezocomposite. The overall conclusion of the above simulations is that the metastructure-based FGM is more suitable for the desired application, not only because of its better RF properties, but also because it offers the required flexibility for an implantable sensor.

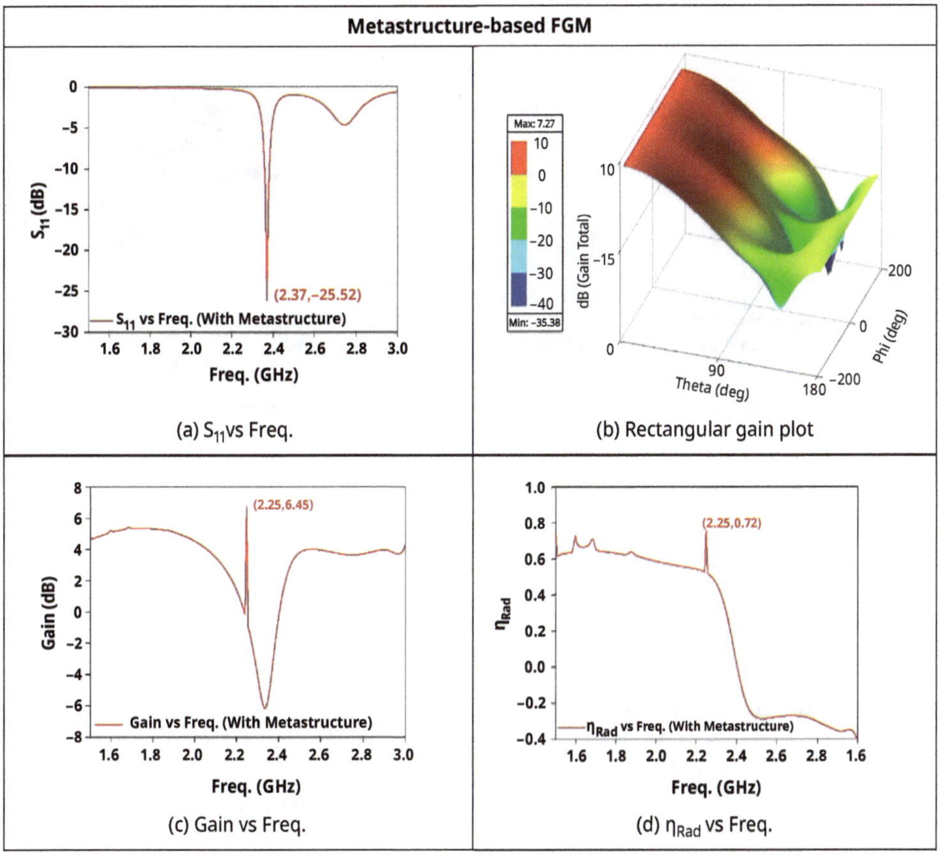

Figure 8.4: R-F characteristics of the sensors with the substrates made up from different FGMs.

8.6.2 Morphological characteristics

The morphological characteristics of the prepared samples were extracted from FESEM images captured at different magnifications, and the samples were also bent to some extent to ensure the substrate's structural stability during deformation (Figure 8.5a and b). It was noticed that the top layer of the gradient-based structure got distorted upon bending, whereas the metastructure provided enough stability and flexibility to the sample for high-loading applications. In the case of metastructure-based FGM, elemental composition was determined by EDS analysis, and the presence of Ca and N confirmed the presence of HAP and CS traces in the 3D-printed structure (Figure 8.5c), thereby highlighting the uniform blending of the nanoparticles in the PVDF-based matrix. Similarly, upon identifying the composition of the gradient-based FGM (Figure 8.5d), it was observed that the nanoparticles of HAP, CS, and $BaTiO_3$ are not detectable on the surface of the substrate sample, whereas PVDF dominates the

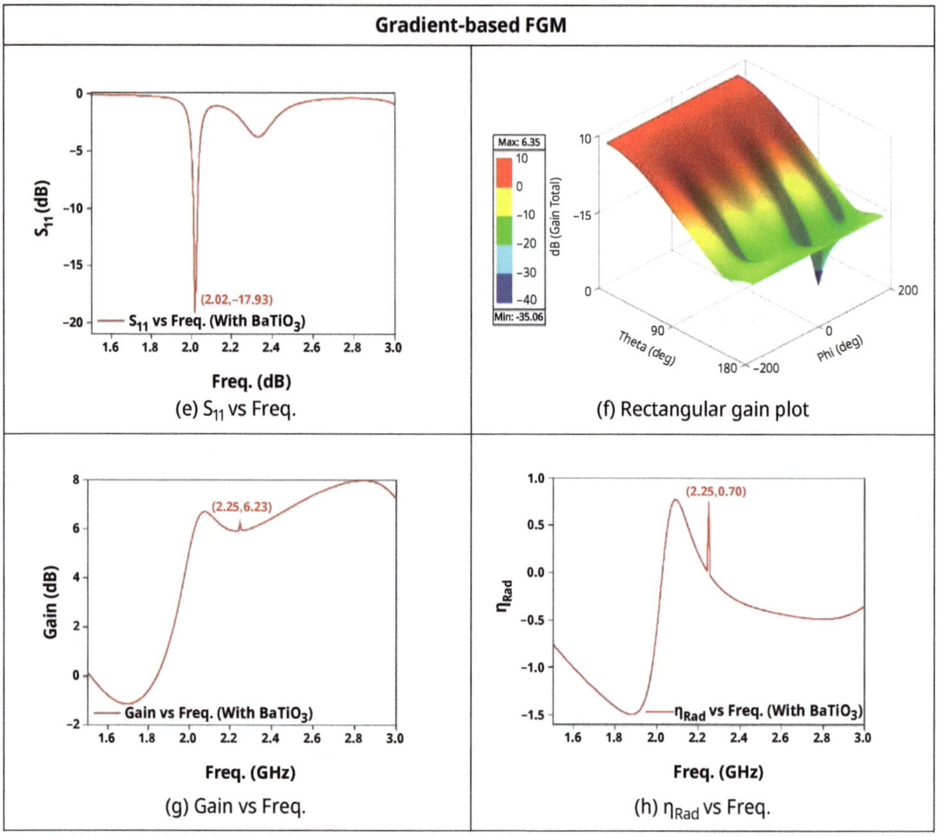

(e) S_{11} vs Freq.

(f) Rectangular gain plot

(g) Gain vs Freq.

(h) η_{Rad} vs Freq.

Figure 8.4 (continued)

surface. The absence of nanoparticles on the surface is due to the lower reinforcement content. The area mapping was also used to localise the presence of different elements within a highlighted area of the sample (Figure 8.5e and f). Following that, the average surface roughness (R_a) of the samples' top surfaces was measured to assess its effect on the sensor's sensing characteristics. It was observed that the metastructure-based sample has a lower surface roughness (95.66 nm) (Figure 8.5g) than the gradient-based sample (100.3 nm) (Figure 8.5h), due to a more uniform dispersion of HAP and CS nanoparticles. The fabrication method also affected the R_a of the gradient-based FGM sample, as printing was carried out in two intervals: first, the two layers of PVDF-HAP-CS were printed, followed by the $BaTiO_3$ layer. To further assess the uniform dispersion of the nanoparticles, 3D-rendered data from the surface profiles were extracted (Figure 8.5i and j), and it was observed that more peak-valley combinations were observed in the gradient-based sample; agglomeration of the particles at certain locations was noted. For the fabrication of this MPA-based sensor, its surface should be smooth enough to minimise return losses during operation; therefore, the metastructure-based FGM provides greater stability and a smoother surface than.

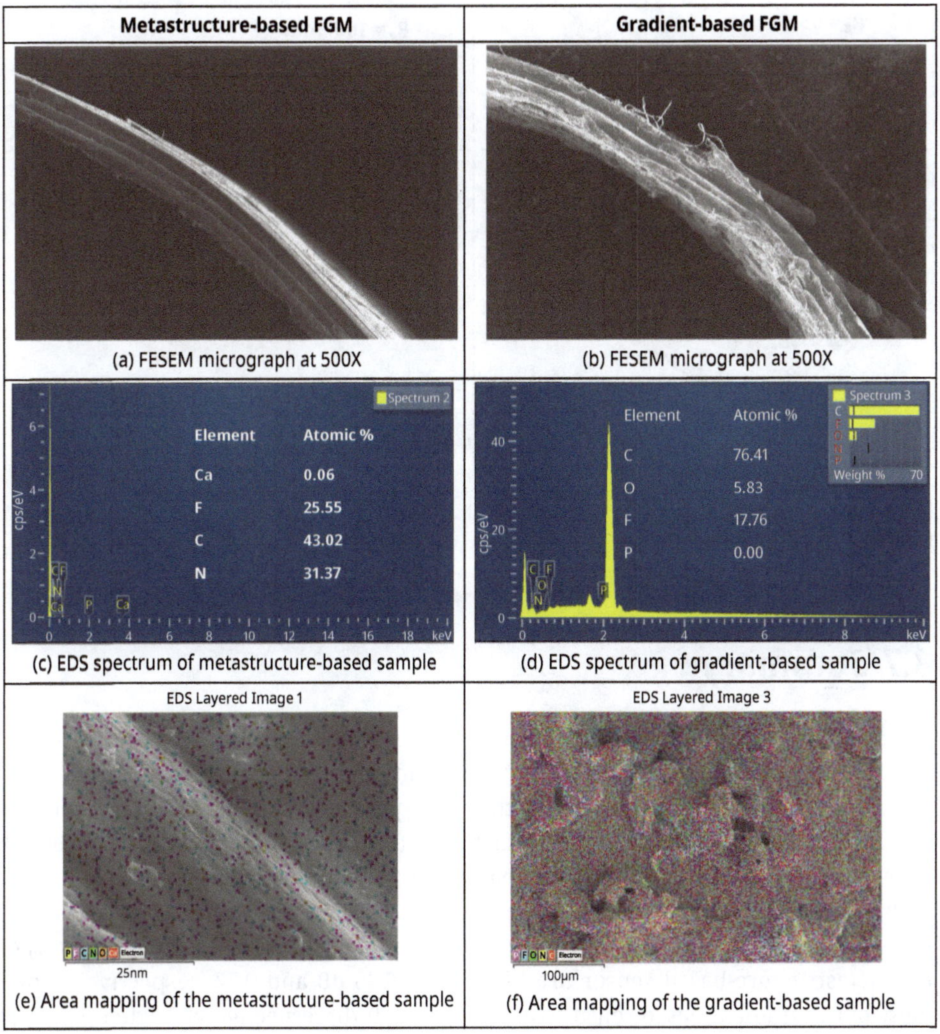

Figure 8.5: Morphological characteristics of the 3D-printed FGM samples.

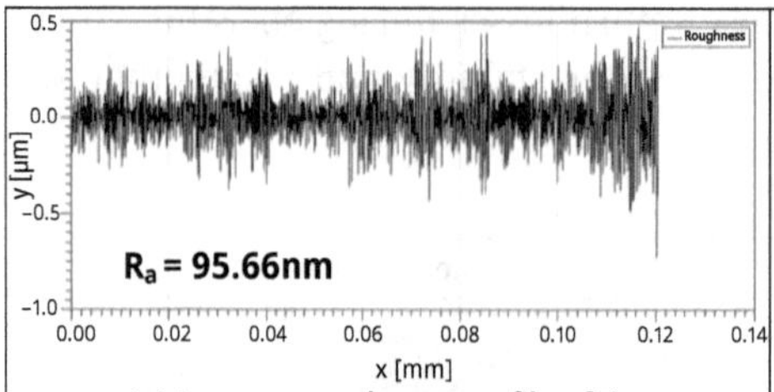

(g) Average roughness profile of the metastructure-based sample

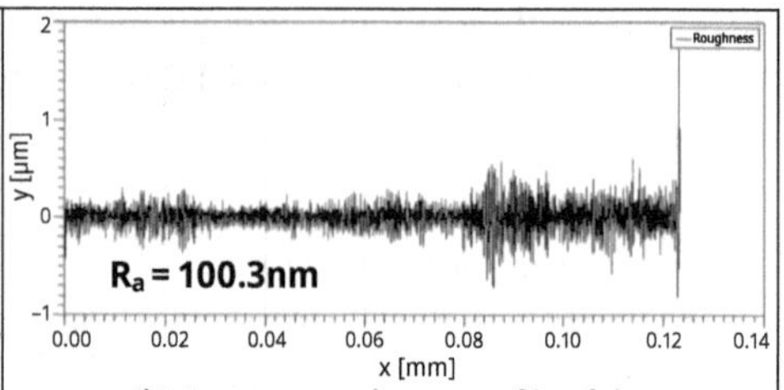

(h) Average roughness profile of the gradient-based sample

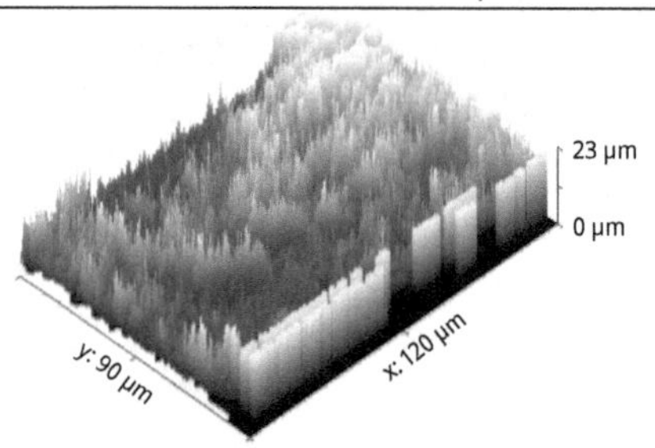

(i) 3D rendered data of the metastructure-based sample

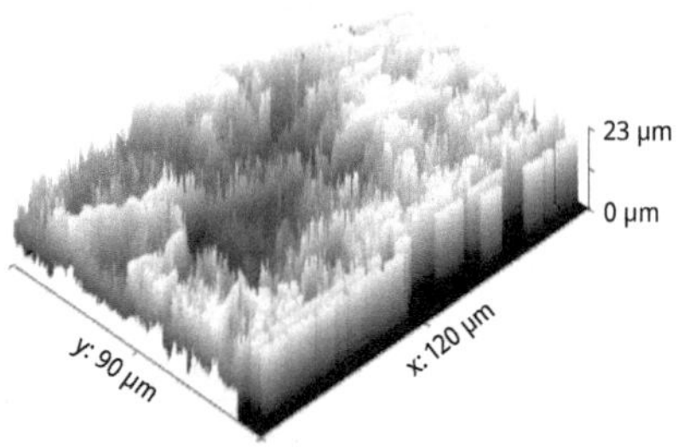

(j) 3D rendered data of the metastructure-based sample

Figure 8.5 (continued)

8.7 Summary

Investigations into the sensing and morphological characteristics of 3D-printed piezoelectric functional structures provide insights into the challenges encountered in designing and fabricating substrates for the two types of FGMs. The design and simulation results show that the sensor with metastructure-based FGM exhibits desirable RF characteristics in the ISM frequency band and resonates very closely (i.e. 2.37 GHz with S_{11} of −25.52 dB) to the targeted frequency (2.45 GHz), whereas the gradient-based sensor resonates at 2.02 GHz and shows an S_{11} of −17.93 dB. Also, the gain and η_{Rad} of the metastructure-based sensor are higher (i.e. 6.45 dB and 0.72, respectively) than those of the gradient-based sensor (i.e. 6.23 dB and 0.70, respectively). Higher gain and efficiency are more desirable for implantable applications to better accommodate signal losses due to body tissues. The morphological characteristics of the prepared samples indicates that the metastructure based substrate possesses more flexibility without any distortion in the layers, whereas, the gradient-based substrate has been undergone some deformation. The element composition results indicated that more uniform dispersion of the nanoparticles was noticed in the first case, whereas, more frequent peak and valley combination was observed in case of gradient-based structure. Also, the average surface roughness profile of the two substrates indicates that the metastructure based substrate exhibits low R_a (95.66 nm) as compared to the gradient-based (100.3 nm), which therefore represents that the first one is more suitable for the sensor fabrication with minimum reflective losses.

References

[1] Iwuozor KO, Jimoh TA, Ojo HT, Emenike EC, Emeghai J, Adeniyi AG. Polymer-based piezoelectric materials: Structure, mechanism, applications, and future trends. Surfaces and Interfaces. 2025;70:106855.

[2] Akter M, Anik HR, Tushar SI, Tania IS, Chowdhury MKH, Hasan SMM, Bristy BF. Advances in functionalized applications of graphene-based wearable sensors in healthcare. Advanced Sensor Research. 2024;3(2):2300120.

[3] Shepelin NA, Glushenkov AM, Lussini VC, Fox PJ, Dicinoski GW, Shapter JG, Ellis AV. New developments in composites, copolymer technologies and processing techniques for flexible fluoropolymer piezoelectric generators for efficient energy harvesting. Energy and Environmental Science. 2019;12(4):1143–76.

[4] Ramadan KS, Sameoto D, Evoy S. A review of piezoelectric polymers as functional materials for electromechanical transducers. Smart Materials and Structures. 2014;23(3):033001.

[5] Suthar V, Joseph J, Kadam S, Bhatt R, Shah B. Functionalized polymer nanocomposites for dielectric applications. In Advances in Functionalized Polymer Nanocomposites 2024 (pp. 915–58). Woodhead Publishing.

[6] Wang G, Qiu W, Liu Y, Bai Y, Cao C, Tang F, ... Zhang M. Damage detection for structural health monitoring using ultra-sensitive flexible piezoelectret sensors. Structural Health Monitoring. 2023;22(4):2800–12.

[7] Yan M, Zhong J, Liu S, Xiao Z, Yuan X, Zhai D, Zhang Y. Flexible pillar-base structured piezocomposite with aligned porosity for piezoelectric energy harvesting. Nano Energy. 2021;88:106278.

[8] Karaca MM, Esen I. On the effect of foam structures on thermo-mechanical vibration response of functionally graded piezoelectric smart plates. International Journal of Mechanics and Materials in Design. 2025;21(4):849–75.

[9] Islak S, Abushraıda AM. Characterization of functionally graded bronze matrix ceramic reinforced composite materials. Sakarya University Journal of Science. 2019;23(6):1137–43.

[10] Panchal Y, K P. Functionally graded materials: A review of computational materials science algorithms, production techniques, and their biomedical applications. Proceedings of the Institution of Mechanical Engineers, Part C: Journal of Mechanical Engineering Science. 2022;236(22):10969–86.

[11] Oshkour AA, Abu Osman NA, Yau YH, Tarlochan F, Wan Abas WAB. Design of new generation femoral prostheses using functionally graded materials: a finite element analysis. Proceedings of the Institution of Mechanical Engineers, Part H: Journal of Engineering in Medicine. 2013;227(1):3–17.

[12] Pandey S, Pradyumna S, Gupta SS. Static and dynamic analyses of functionally graded sandwich skew shell panels. Journal of Sandwich Structures and Materials. 2021;23(8):4135–69.

[13] Ercument DB, Sahmani S, Safaei B. A review on computational linear and nonlinear dynamic analysis of shell-type composite structures. Computers and Structures. 2025;306:107596.

[14] Pehlivan F, Esen I, Aktas KG. Thermomechanical response of smart magneto-electro-elastic FGM nanosensor beams with intended porosity. Arabian Journal for Science and Engineering. 2025;50 (4):2755–77.

[15] Wang H, Zhai X, Xu J, Yang L. Effect of sintering time on structure and properties in CuO-doping KNN-LS-BF piezoelectric ceramics. Journal of Wuhan University of Technology-Materials Science Edition. 2019;34(2):308–11.
[16] Bodkhe S, Ermanni P. Challenges in 3D printing of piezoelectric materials. Multifunctional Materials. 2019;2(2):022001.
[17] Hamza M, Kanwal Q, Hussain MI, Khan K, Asghar A, Liu Z, Chen Z. Recent progress in 3D printed piezoelectric materials for biomedical applications. Materials Science and Engineering: R: Reports. 2025;164:100962.
[18] Chorsi MT, Curry EJ, Chorsi HT, Das R, Baroody J, Purohit PK, Nguyen TD. Piezoelectric biomaterials for sensors and actuators. Advanced Materials. 2019;31(1):1802084.
[19] Bouad V, Girardot M, Ladmiral V, Barrau S. Piezoelectric fluorinated polymer composites: a review on coupling agents at the filler/matrix interface. Polymer Composites. 2024;45(5):3861–82.
[20] Joseph J, Kumar M, Tripathy S, Kumar GS, Singh SG, Vaniari SRK. A highly flexible tactile sensor with self-poled electrospun PVDF nanofiber. In 2018 IEEE SENSORS 2018, October (pp. 1–4). IEEE.
[21] Pourmadadi M, Ahmari A, Mirshafiei M, Omrani Z, Yazdian F, Rahdar A, Aboudzadeh MA. Polyvinylidene fluoride in biomedical applications: Properties, challenges, and future prospects. European Polymer Journal. 2025;231:113889.
[22] Ranjan N, Singh R, Ahuja IPS, Rahman M, Ramakrishna S. PLA-HAp-CS-based biocompatible scaffolds prepared through micro-additive manufacturing: a review and future applications. 3D printing in Biomedical Engineering. 2020;209–29.
[23] Singh S, Singh G, Prakash C, Ramakrishna S, Lamberti L, Pruncu CI. 3D printed biodegradable composites: An insight into mechanical properties of PLA/chitosan scaffold. Polymer Testing. 2020;89:106722.
[24] Li S, Shan Y, Chen J, Chen X, Shi Z, Zhao L, Li Y. 3D printing and biomedical applications of piezoelectric composites: A critical review. Advanced Materials Technologies. 2025;10(5):2401160.
[25] Curry EJ, Le TT, Das R, Ke K, Santorella EM, Paul D, Nguyen TD. Biodegradable nanofiber-based piezoelectric transducer. Proceedings of the National Academy of Sciences. 2020;117(1):214–20.
[26] Zhou Y, Chen CC, Volakis JL. Dual band proximity-fed stacked patch antenna for tri-band GPS applications. IEEE Transactions on Antennas and Propagation. 2007;55(1):220–23.
[27] Barwar A, Kala P, Singh R. 3D-printed conformal sensors for health monitoring of bovine post diaphragmatic hernia surgery: a review. Rapid Prototyping Journal. 2024;30(5):904–18.
[28] Singh R, Kumar S, Singh AP, Wei Y. On comparison of recycled LDPE and LDPE–bakelite composite based 3D printed patch antenna. Proceedings of the Institution of Mechanical Engineers, Part L: Journal of Materials: Design and Applications. 2022;236(4):842–56.
[29] Singh G, Singh R, Singh AP, Anand A, Kumar V. On Novel 3D-Printed Polylactic Acid Composite Matrix for Condition Monitoring of Comminuted Fractures. Journal of Materials Engineering and Performance. 2025;34(19):21456–64.

Vinay Kumar*, Ranvijay Kumar, Rupinder Singh

Chapter 9
Shape memory and 4D capabilities of the 3D-printed piezoelectric materials

Abstract: Studies on shape-memory polymers over the past two decades have highlighted the usefulness of these piezoelectric materials in the manufacturing of smart sensors and actuators for robotics, aerospace, and automotive engineering. Shape-memory properties based on the 4D printing capabilities of different thermoplastic composites have been widely reported for tactile and pressure-sensing applications. But hitherto, very few studies have been reported on one- and two-way programmable shape-memory thermoplastic composites with 4D printing properties for biosensing applications. This chapter highlights investigations into the shape-memory and 4D capabilities of in-house-fabricated piezo-responsive polyvinylidene fluoride and polylactic acid thermoplastic composites. The acceptable shape-recovery properties of a 3D-printed piezoelectric functional prototype indicate that the proposed composite structures with tunable features may be successfully used for the 3D printing of customized scaffolds, implants, and drug-delivery solutions for biomedical applications. Such characteristics of 3D-printed piezoelectric materials may be used to manufacture adaptive wings and auxetic structures for aerospace vehicles. Also, the self-expanding and self-contracting (shape-memory) features of such prototypes may be used in self-assembling robots and in the preparation of soft grippers for robots. The shape-memory properties of highly flexible 3D-printed piezoelectric composite may be used for wearable electronics manufacturing, digital twin solutions for mechanical systems, and condition health monitoring of built structures.

Keywords: Thermoplastics, PVDF, piezoelectric composites, shape memory, 4D printing

9.1 Introduction

The experimental investigation reported in the past two decades on electroactive, temperature-dependent, and magnetostrictive shape-memory thermoplastic materials

***Corresponding author: Vinay Kumar,** Mechanical Engineering Department, Marwadi University Research Center, Marwadi University, Rajkot, Gujarat, India, e-mail: vinaykumar6192@gmail.com
Ranvijay Kumar, Mechanical Engineering Department, Marwadi University Research Center, Marwadi University, Rajkot, Gujarat, India
Rupinder Singh, National Institute of Technical Teachers Training and Research, Chandigarh, India

 | https://doi.org/10.1515/9783112218846-009

has outlined that piezoelectric properties (i.e., electro-shape-memory-assisted quick actuation) play a significant role in the design and development of novel pressure and tactile sensors as well as the manufacturing of nonlinear actuators [1]. Shape-memory materials with piezoelectric actuation properties have reportedly been used to monitor small changes in energy consumption and voltage operation of electrical equipment using pulsed voltage. The polarization capabilities of piezoelectric materials have been recognized as an important factor in imparting acceptable shape-memory properties in the prepared material for such applications [2]. It has been reported that the poling-assisted electrostrictive properties of piezoelectric composites impart a high piezoelectric coefficient to the fabricated component for self-sensing and self-actuator applications [3]. Based on such a significant rise in studies reported on shape-memory polymers for 4D printing, Figure 9.1 shows the increase in the trends for studies reported (i.e., a total of 899 documents as per the "Scopus" database) on 4D-printable shape-memory polymers with smart piezoelectric properties in the past decade.

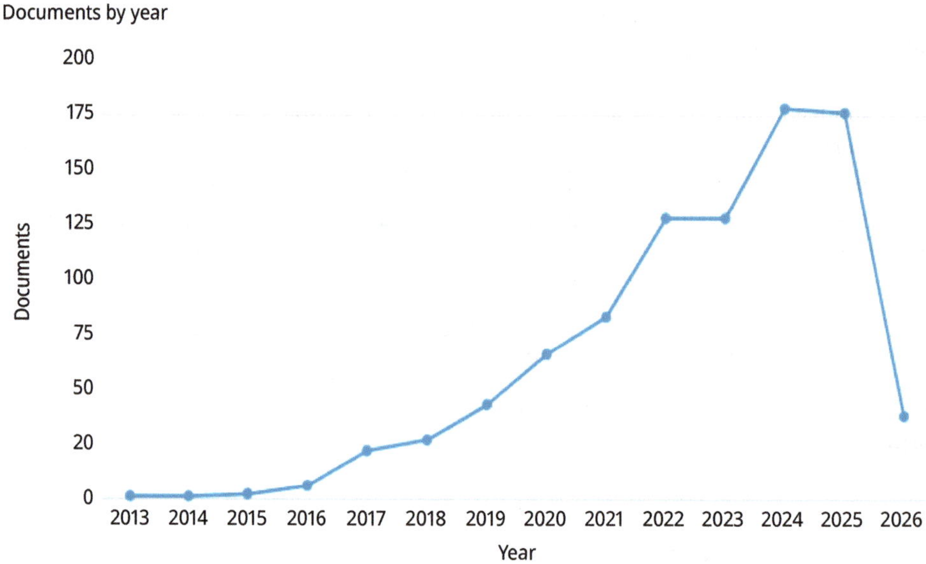

Figure 9.1: Studies reported in the past decade on shape-memory polymers for 4D printing applications.

The design and fabrication of metal alloy wire-based composite structures with piezoelectric behavior have been reported recently by some researchers for the manufacture of smart shock absorbers. It has been experimentally demonstrated that the composition of five metallic alloys exhibits a highly acceptable shape-memory effect for a smart, controlled damping effect. Such piezoelectric composite structures may be used for aerospace exploration practices [4]. The review studies reported on the comparison of shape-memory materials (prepared from metal alloys and polymer composite matrices) and piezoelectric composite materials and indicated that the smart materials with such features have a great scope of applications in biosensors and health monitoring solutions. The electromechanical response of shape-memory alloys, as well as the electrical poling-assisted dipole moment in piezoelectric composites, may be used to fabricate smart consumable products for automobiles, robots, structural health monitoring, and biomedical surgical practices [5]. In contrast to such review literature, the rise in studies on smart 4D materials with shape-memory properties across various reputable publication sources is shown in Figure 9.2.

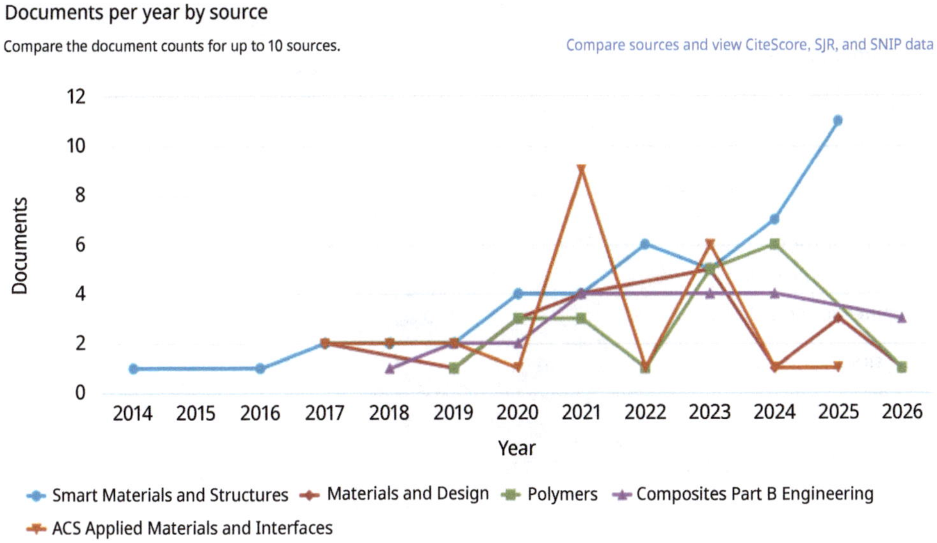

Figure 9.2: The trends observed in the past decade on smart 4D materials with shape-memory properties in various reputed publication sources.

The investigation performed on thermoplastic composites prepared using ionic ceramic-based reinforcements such as lead zirconate titanate, barium zirconate titanate, barium titanate ($BaTiO_3$), zinc oxide (ZnO) has highlighted the potential in such multifunctional smart materials for energy harvesting, biomedical sensors, and engineering and medical applications [6]. The advancement in utilization of smart polymer composite materials (with piezoelectric properties-based shape memory) toward 3D printing technology-based manufacturing of self-powered biomedical solutions like shape-memory scaffolds, and one-way programmable implants have highlighted the integrated advantages advanced manufacturing practices with smart materials for customized solution preparation [7]. Four-dimensional printing has been outlined by the researchers as a highly promising domain in the present time to use the smart piezoelectric materials with shape-memory properties for variety of medical applications and programmed actuation system with interdigital electrode assembly [8, 9]. The advantages of 4D-printed biocompatible smart materials, such as polyvinylidene fluoride (PVDF) and polylactic acid (PLA), have been widely reported for dentistry and shape-morphing applications. This shows that polymer matrices with tunable properties have a wide scope in surgical practices for both human and veterinary animals [10, 11]. The significant contributions of subject areas such as materials science and physics to the design and development of piezoelectric, semiconductor-based smart composites underscore the potential for research and development of such materials for 4D printing applications. Figure 9.3 highlights the distribution of reported literature by subject area for shape-memory-assisted 4D-printable materials.

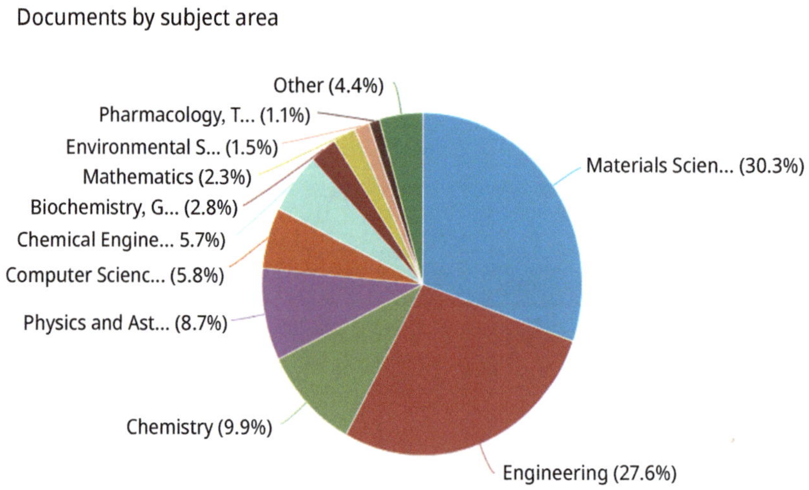

Figure 9.3: Pie chart distributions for subject area-wise reported literature on shape-memory-assisted 4D-printable materials.

The preparation of polyethylene terephthalate-based smart hydrogels has been reported by some researchers. Such smart hydrogel materials may be used to manufacture 4D-printed scaffolds as novel biomedical solutions with orthogonal photo-reactions [12]. The review articles reported on various piezoelectric materials that may be processed to fabricate novel engineering devices using different additive manufacturing practices and have thoroughly outlined the advantages, difficulties, and future scope of 3D- and 4D printable smart piezoelectric composites for bioengineering applications [13, 14]. Further, Figure 9.4 outlined the categorization of literature on shape-memory polymers by document type, including review papers, articles, book chapters, etc. The categorization indicated that 70.4% of the studies reporting on shape-memory-capable 4D materials have been published as research articles.

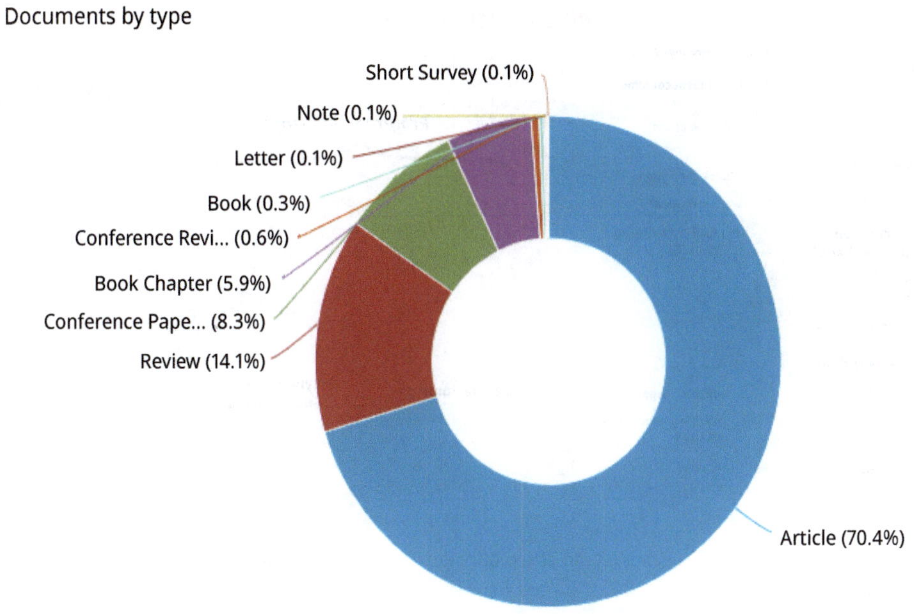

Figure 9.4: Distribution for the type of documents reported on shape-memory-assisted 4D-printable polymer matrix for various engineering applications.

The role of advanced 3D/4D printing has been reported in the development of sustainable civil engineering solutions. The porous, meta-structures, and auxetic properties-based smart design of civil structures and their 4D printing have been outlined as the future of civil engineering solutions to construct more energy-efficient, airy, quake-proof, spacious, and defect-proof smart civil structures. These models may be used to develop cost-effective schools, houses, and office spaces in remote locations, as well as to boost sustainable growth across different geographical locations [15]. Some researchers have recently reported the preparation of semicrystalline polymers with shape-memory features, assisted self-healing in 3D-printed spiral structures, and compressed and folded prototypes. This outlined that 4D printable self-repair or healing scaffolds and implants may be used for smart bioengineering applications [16, 17]. Figure 9.5 shows the activation mechanism of shape-memory properties in semicrystalline polymers that may be used for 4D printing applications.

The 4D printing capabilities of thermoplastic composite blends/blends, such as ZnO-reinforced PLA and polycaprolactone, are being researched for biomedical scaffolds and implant fabrication. The prepared solution may be used in human and veterinary surgical practices [18, 19]. Regarding tissue engineering, novel investigations have reported 4D-printable bio-piezoelectric smart scaffolding materials that also exhibit good flexibility for tactile sensing applications. Such smart sensors, as well as biomedical solutions, have also been found useful for online health condition moni-

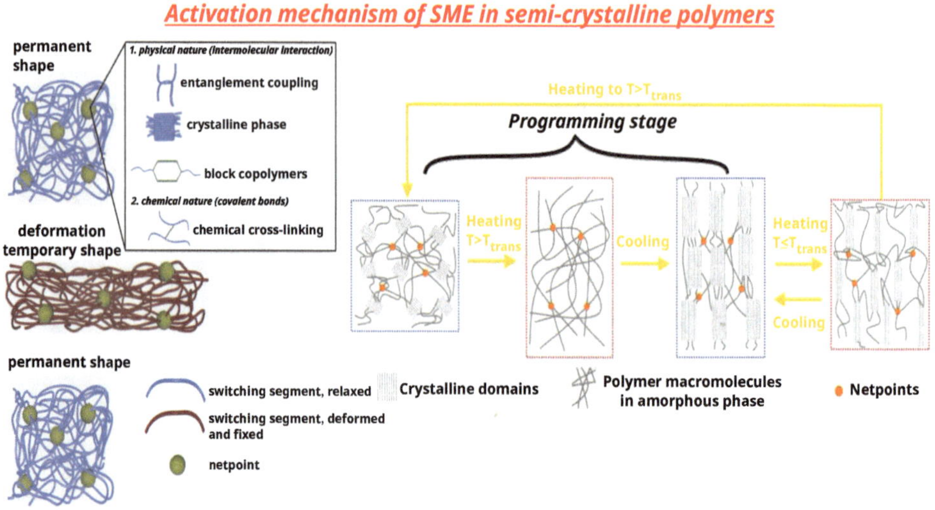

Figure 9.5: Activation mechanism scenario schematics in shape-memory features based semicrystalline polymers [16].

toring of patients after diaphragmatic hernia, hip, and bone joint surgeries [20, 21]. The studies reported on piezoelectric PVDF composites and its ferroelectric trifluoro ethylene (TrFE)-based PVDF-TrFE matrix have indicated that tuning the crystallization behavior of these matrices plays a significant role in imparting high electroactive properties in the 4D-printed composite structures for highly flexible, wearable, electronic devices, and modern condition monitoring sensors [22–24]. Such studies reported on the advances of smart piezoelectric and shape-recovery polymer composite matrix toward 3D/4D and 5D printing applications and described that such crystalline matrices with highly acceptable electrostrictive and magnetostrictive features may be used effectively in transducing the applied load to electrical signals and vice versa. The output obtained from 3D-printed smart piezoelectric composites may be used to fabricate shape-recovery-based 4D behavior in one- and two-way programmable sensors, as well as in augmented reality/virtual reality-assisted smart digital twinning solutions [25, 26].

9.2 Research gap and problem formulation

The literature review revealed that significant studies have been reported on the in-house fabrication of smart piezoelectric materials with shape-memory-based programmable features for 4D printing applications. The Scopus database of 899 reported documents was processed in the VOSviewer software package to investigate the core domains in which studies on 4D printing have been explored to date. It was noted that a total of 20 terms related to design, reinforcement/filler, stimulus, programming, and shape recovery, among others, of electroactive piezoelectric polymer matrices were identified to yield 262 highly relevant keywords covering material properties and their innovative applications. Based on these relevant keywords, 60% most relevant key terms (equivalent to 155 terms) created a web of reported terms. This web of keywords, interconnecting the studies reported on various approaches to smart material development and their applications in bioengineering, is outlined in Figure 9.6.

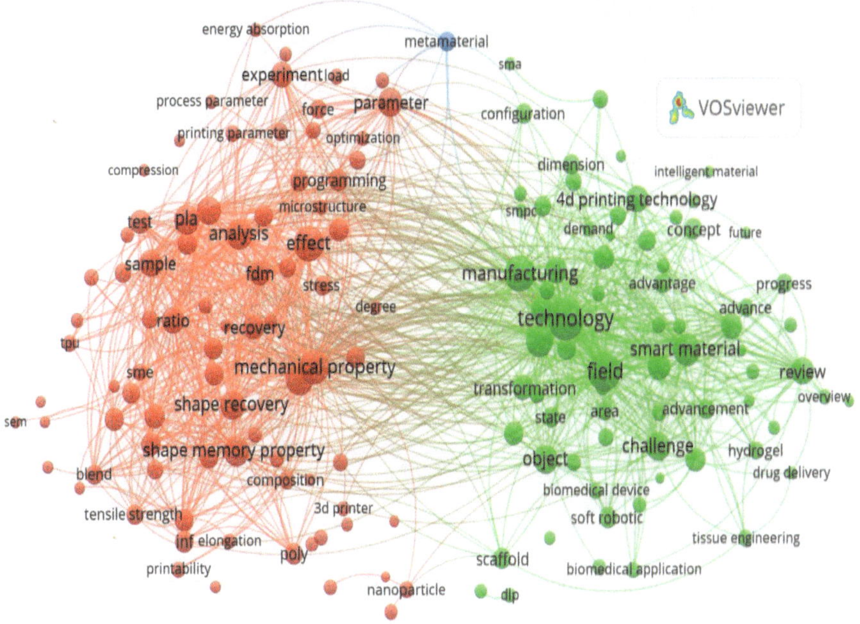

Figure 9.6: Web of key reported terms on preparation of shape-memory polymers for 4D printing applications.

Based on this web, it may be stated that the investigations reported on the shape memory and 4D capabilities of smart piezoelectric materials primarily focus on two major domains of research: the design and development of smart materials and the exploration of suitable application areas for such materials. The interconnection of these two research domains, outlined in a web of keywords, was further explored to identify the gap in the literature. Figure 9.6 (with the relevant node on shape-memory polymers) shows that the web of studies on the development of such smart materials does not always aim for 4D printing. Also, the preparation of shape-memory polymers is mainly focused on the methods of manufacturing novel products from smart materials. Also, Figure 9.7 indicates that studies on 4D printing technology have focused on the use of acceptable programmability features and rheological, thermal, and mechanical properties for manufacturing novel bioengineering solutions. As per the research gap analysis, Figures 9.8 and 9.9 outline the gap in research: although shape-memory properties based on 4D printing capabilities of different thermoplastic composites have been widely reported for tactile and pressure sensing applications. But very few studies have reported on one- and two-way programmable shape-memory thermoplastic composites with 4D printing properties for biosensing applications.

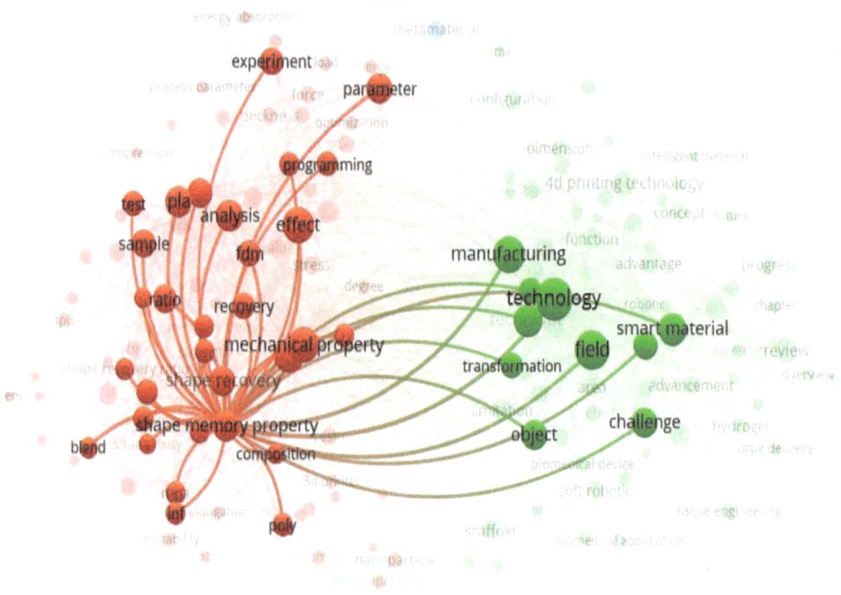

Figure 9.7: Web of keywords based on studies associated with shape-memory polymers.

In this study, the process capability of a 4D-printable polymer-matrix-based composite prepared from PVDF has been investigated for acceptable shape-memory-assisted piezoelectric sensing applications.

9.3 Case study for the shape-memory analysis and 4D capabilities of the piezoelectric functional structures

The 4D printing and shape-memory performances depend heavily on the materials and design considerations. Also, in some specific applications, the 4D capabilities of manufactured parts are strongly influenced by their designs. In fact, in sensing

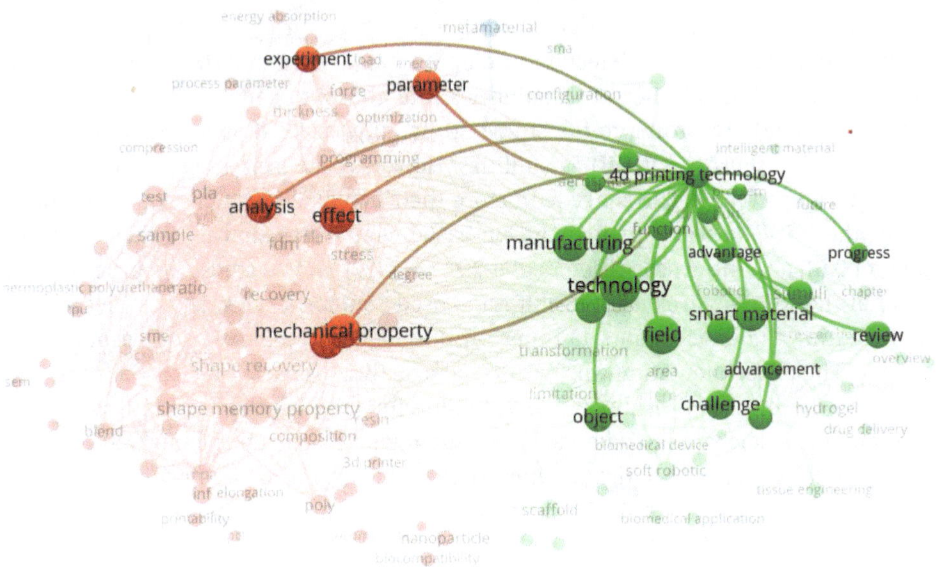

Figure 9.8: Web of terms reported on 4D printing technology of smart materials.

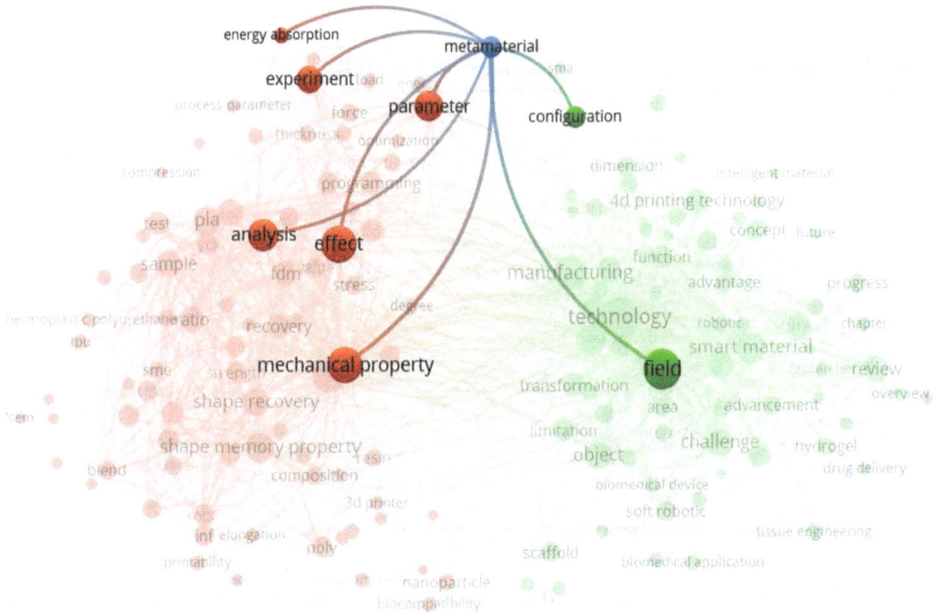

Figure 9.9: Gap in the literature outlining the missing interconnection of shape-memory polymers for 4D printing applications.

applications, the design of the antenna and sensors is defined by the dielectric constant. Barwar et al. [27] investigated the design and fabrication (3D printing) of a PVDF-based sensor for the detection of diaphragmatic hernia. PVDF is one of the established materials that has 4D capabilities (the ability to generate electricity when force is applied). According to the study, 3D printing of PVDF is a potential solution for developing sensors with 4D capabilities. However, the process capability analysis of the developed materials is needed before their actual applications. In the reported study, the PVDF-hydroxyapatite-chitosan-based composites showed a dielectric constant of 9.61, and the sensor resonated at 2.72 GHz [27]. In this study, the process capability of the PVDF-hydroxyapatite-chitosan-based composite sensors was investigated by repeating the experiments 15 times on the dielectric constant and resonant frequency. The experimentation was performed exactly as reported by Barwar et al. [27]. Based on the previous results, the target value for the dielectric constant was set at 9.61, and considering a 5% data variation, the lower specification limit (LSL) was set at 9.125 and the upper specification limit (USL) at 10.0905. The target value for the resonant frequency was set at 9.72 GHz, and considering a 5% variation in the data, the LSL was set at 2.584 GHz and the USL at 2.856 GHz. Table 9.1 shows the dielectric constant and resonant frequency values from the repetitive experiments.

Table 9.1: Dielectric constant and resonating frequency at repetitive experiments.

Sample	Dielectric constant	Resonating frequency (GHz)
1	9.65	2.71
2	9.59	2.70
3	9.75	2.77
4	9.60	2.67
5	9.58	2.69
6	9.77	2.78
7	9.80	2.68
8	9.48	2.68
9	9.67	2.71
10	9.64	2.75
11	9.87	2.72
12	9.74	2.67
13	9.55	2.72
14	9.66	2.69
15	9.82	2.73

Figure 9.10 shows the normal distribution plot for the dielectric constant of the PVSD-based sensor.

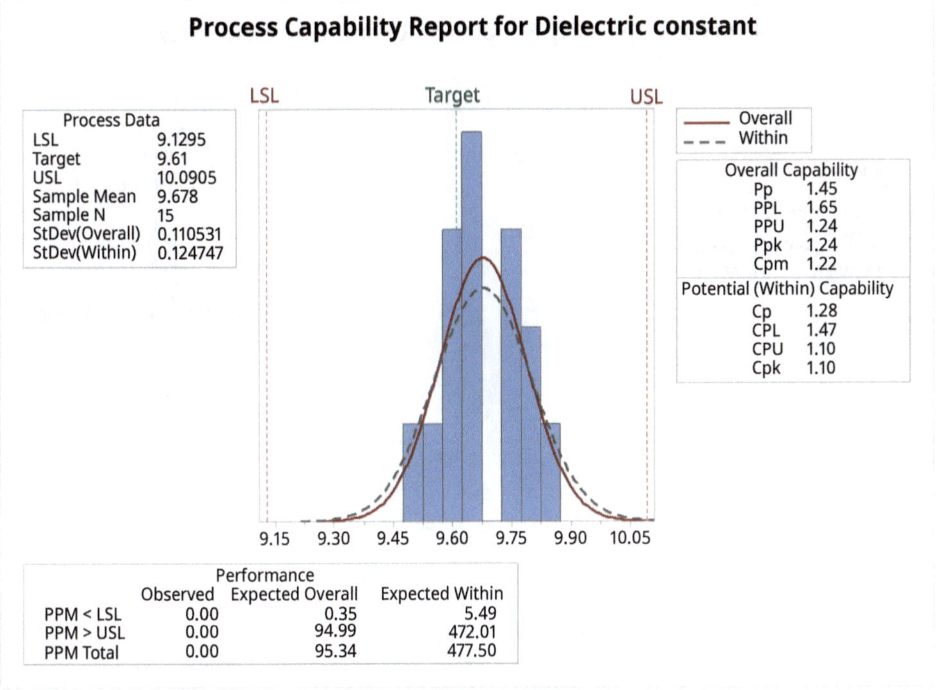

Figure 9.10: Process capability indicators for dielectric constant.

It was observed that the values were consistent and close to the target value most of the time, with the overall standard deviation of 0.110. Along with these, the C_p was 1.28, and the C_{pk} was 1.10, both greater than 1, indicating good process capability for the given data. In fact, the results for the dielectric constant indicate that 3D printing of the PVDF-based composite materials is capable of precision manufacturing (Figure 9.9).

Similarly, the analysis also suggests a good hold on process capability for the values of the resonating frequency. Process capability indicators for resonating frequency showed that 3D printing alone is not sufficient; the sensor design is also correct. The observed values are near the target value most of the time, with an overall standard deviation of 0.0344. Along with these, the C_p was observed to be 1.29 and the C_{pk} to be 1.21, both greater than 1 (Figure 9.11), indicating good process capability. The 3D printing of the PVDF-based composite materials and the design are acceptable and meet the process capability parameters.

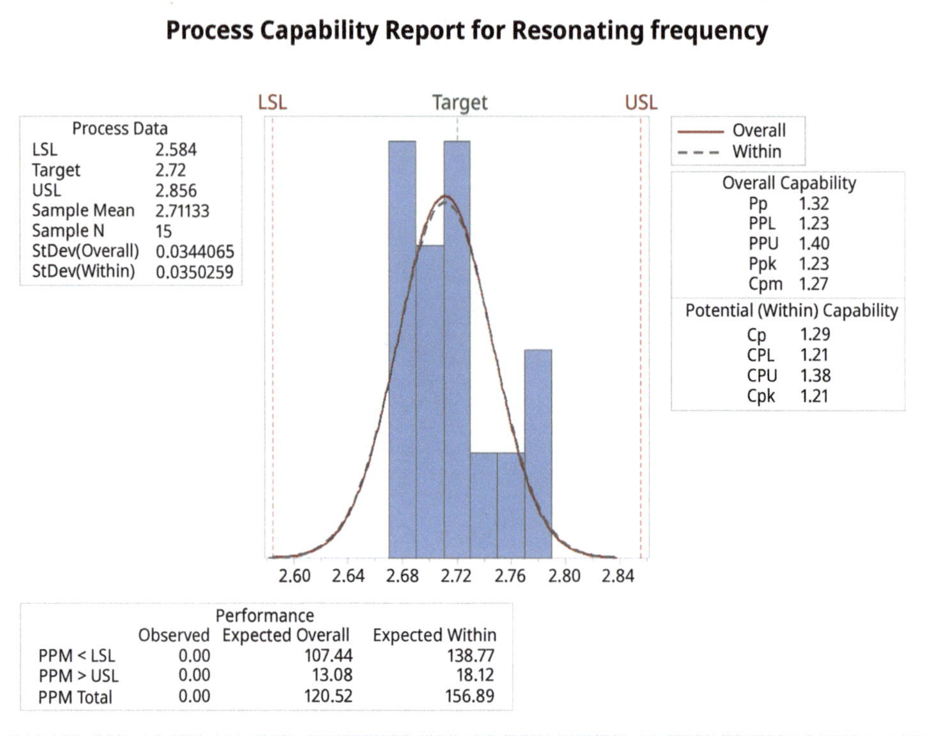

Figure 9.11: Process capability indicators for resonating frequency.

9.4 Conclusions

The investigations on dielectric property-based assessment of one- and two-way programmable thermoplastic composites of PVDF-hydroxyapatite-chitosan for piezoelectric biosensing practices highlighted the potential of the proposed composition for the detection of diaphragmatic hernia recurrence in veterinary animals. The results of the process capability analysis of dielectric constant for 4D-printable PVDF composite samples indicated that the process has good capability, with $C_p = 1.28$ and $C_{pk} = 1.10$ (both desirable factors greater than 1), which is highly acceptable for fabricating the proposed composition-based piezoelectric sensors for engineering and biomedical applications. Similarly, good process capability was observed for the resonating frequency, with $C_p = 1.29$ and $C_{pk} = 1.21$ (both >1), indicating that the PVDF composite is acceptable for 4D-printed piezoelectric sensing applications. The 3D-printed sensor based on a PVDF-based composite matrix with a tunable dielectric constant can be effectively utilized for the design and development of conformal and implantable biosensors. The shape memory and 4D capabilities can assist in altering the dielectric constant of a 3D-printed functional prototype/piezoelectric biosensor, enabling tuning of the resonant frequency (R_f) signal output for effective online health monitoring of treated patients after surgical procedures.

References

[1] Ren X, Zhang LX. Electro-shape-memory effect in ferroelectric martensite. Materials Science and Engineering: A. 2006 Nov 25;438:1071–76.

[2] Morita T, Kadota Y, Hosaka H. Shape memory piezoelectric actuator. Applied Physics Letters. 2007 Feb 19;90(8).

[3] Kadota Y, Hosaka H, Morita T. Shape memory piezoelectric actuator by control of the imprint electrical field. Ferroelectrics. 2008 Oct 21;368(1):185–93.

[4] Liu M, Xu J, Li Q. Design and experiment of piezoelectric-shape memory alloy composite shock absorber. Materials Letters. 2021a Dec 1;304:130538.

[5] Chatterjee A. Critical Comparison of Smart Materials: Shape Memory Alloys vs Piezoelectric Materials: A Thorough Review. Acta Scientific Applied Physics Volume. 2023 Mar;3(3).

[6] Zhao F, Zhou B, Zhu X, Wang H. Constitutive model of piezoelectric/shape memory polymer composite. In Journal of Physics: Conference Series 2024 Feb 1 (Vol. 2713, No. 1, p. 012037). IOP Publishing.

[7] Li B, Ma Y, Fatima K, Zhou X, Gu X, Chen S, He C. 3D printed shape-memory piezoelectric scaffolds with in-situ self-power properties for bone defect repair. Journal of Nanobiotechnology. 2025a Mar 24;23(1):244. doi: 10.1186/s12951-025-03325-x. PMID: 40128753; PMCID: PMC11934793.

[8] Grinberg D, Siddique S, Le MQ, Liang R, Capsal JF, Cottinet PJ. 4D Printing-based piezoelectric composite for medical applications. Journal of Polymer Science Part B: Polymer Physics. 2019 Jan 15;57(2):109–15.

[9] Li J, Yan M, Zhao L, Zhai D, Zhong C, Luo H, Zhang Y, Yuan X, Zhang D. 4D printed piezoelectric ceramics with patterned interdigital electrodes for actuation application. Ceramics International. 2025b May 14.

[10] Aafreen M M, Cholan PK, Ilango P, Parthasarathy H, Tadepalli A, Ramachandran L. Exploring the 4D printing linked bio-smart materials in dentistry: a concise overview. Frontiers in Dental Medicine. 2025 Apr 16;6:1558382.

[11] Georgantzinos SK, Kostopoulos G, Stamoulis KP. Characterizing the shape-morphing behaviour of 4D printed PLA structures. In Journal of Physics: Conference Series 2024 Mar 1 (Vol. 2716, No. 1, p. 012034). IOP Publishing.

[12] Jee C, Matsumoto H, Horiuchi T, Liu Z, Wang Z, Kakeru O, Kojio K, Nagao M, Miura Y. Preparation of 4D hydrogels with PET-RAFT and orthogonal photo-reactions. RSC Applied Polymers. 2025;3 (1):156–62.

[13] Gong G, You Y, Shen H, Laghaei M, Wang Y, Li Y. 3D printing piezoelectric materials: Innovations, challenges, and future perspectives. Materials Today Sustainability. 2025 Nov 19;101257.

[14] Kumar A, Kumar P, Mittal RK, Gambhir V. Materials processed by additive manufacturing techniques. In Advances in additive manufacturing 2023 Jan 1 (pp. 217–33). Elsevier.

[15] Firoozi AA, Firoozi AA. A systematic review of the role of 4D printing in sustainable civil engineering solutions. Heliyon. 2023 Oct 1;9(10).

[16] Dayyoub, T., Maksimkin, A. V., Filippova, O. V., Tcherdyntsev, V. V., &Telyshev, D. V. Shape Memory Polymers as Smart Materials: A Review. Polymers. 2022;14(17):3511.

[17] Zhao W, Yue C, Liu L, Liu Y, Leng J. Research progress of shape memory polymer and 4D printing in biomedical application. Advanced Healthcare Materials. 2023 Jun;12(16):2201975.
[18] Kumar R, Singh R, Kumar V, Kumar P. On PLA–ZnO composite matrix for shape memory effect. In 4D Printing 2022 Jan 1 (pp. 147–60). Elsevier.
[19] Kumar R, Singh R, Kumar V. On 4D Capabilities of Polycaprolactone Composite Matrix. National Academy of Sciences Letters. 2025 Jan 16;1–4.
[20] Chang SM, Hur S, Park J, Lee DG, Shin J, Kim HS, Song SE, Baik JM, Kim M, Song HC, Kang CY. Optimization of piezoelectric polymer composites and 3D printing parameters for flexible tactile sensors. Additive Manufacturing. 2023 Apr 5;67:103470.
[21] Chen A, Su J, Li Y, Zhang H, Shi Y, Yan C, Lu J. 3D/4D printed bio-piezoelectric smart scaffolds for next-generation bone tissue engineering. International Journal of Extreme Manufacturing. 2023 Jun 27;5(3):032007.
[22] Bai X, Li H, Chu X, Liu M, Li H, Wang H, Sun X, Yan S. The tuning of crystallization behavior of ferroelectric poly (vinylidene fluoride-co-trifluoroethylene). Journal of Polymer Science. 2024 May 1;62(9):1742–70.
[23] Jia N, He Q, Sun J, Xia G, Song R. Crystallization behavior and electroactive properties of PVDF, P (VDF-TrFE), and their blend films. Polymer Testing. 2017 Feb 1;57:302–06.
[24] Zhang J, Wang J, Zhong C, Zhang Y, Qiu Y, Qin L. Flexible electronics: advancements and applications of flexible piezoelectric composites in modern sensing technologies. Micromachines. 2024 Jul 30;15(8):982.
[25] Liu K, Zhang Q, Zhou C, Shi Y, Sun C, Sun H, Yin C, Hu J, Zhou S, Zhang Y, Fu Y. 4D printing of lead zirconate titanate piezoelectric composites transducer based on direct ink writing. Frontiers in Materials. 2021b May 28;8:659441.
[26] Megdich A, Habibi M, Laperrière L. Review of progress in 4D printing of piezoelectric energy harvesters. Materials Science and Engineering: R: Reports. 2025 Sep 1;166:101072.
[27] Barwar A, Kala P, Singh R. On the design and fabrication of polyvinylidene fluoride-based sensors to detect recurrence of diaphragmatic hernia. Proceedings of the Institution of Mechanical Engineers, Part C: Journal of Mechanical Engineering Science. 2025 Apr;239(8):2681–99.

Nishant Ranjan*, Rupinder Singh

Chapter 10
Investigations of the 3D printing process parameters on the performance of the piezoelectric structures

Abstract: Additive manufacturing (AM) or three-dimensional (3D) printing technology is one of the most demanding and fast-growing fabrication/manufacturing technologies in different areas (biomedical, piezoelectric, sensing, aerospace, automobile, etc.) due to its higher accuracy, ease of design, and single product production, which makes it popular in the manufacturing industry. It is used to develop innovative functional materials, such as piezoelectric structures for sensing, actuation, and energy harvesting. The productivity and quality of the produced product, particularly piezoelectric components made with AM, are significantly affected by optimizing 3D printing process parameters. This has an immediate impact on the microstructure, mechanical qualities, thermal properties, and electromechanical response of the fabricated end products. This chapter provides a comprehensive analysis of the impact of critical 3D printing process parameters, including printing temperature, layer height, print speed, infill density, and extrusion rate, on the structural and functional properties of piezoelectric final products manufactured using AM technology. Changes in these parameters significantly affect interlayer adhesion, porosity distribution, and particle alignment within the printed matrix, ultimately influencing piezoelectric output and mechanical properties. Microstructural characterization and electrical tests were performed to determine the correlation between process-induced morphological changes and piezoelectric coefficients. Based on this review, it has been observed that specific process parameters in 3D printing are more effective for developing high-performance AM piezoelectric devices for advanced smart-material and sensor applications.

Keywords: Piezoelectric, additive manufacturing, polymer, composite, piezoelectric materials, extrusion

***Corresponding author: Nishant Ranjan,** Marwadi University Research Center, Faculty of Engineering and Technology, Department of Mechanical Engineering, Marwadi University, Rajkot 360003, Gujarat, India, e-mail: ranjan_nishant92@hotmail.com
Rupinder Singh, Mechanical Engineering Department, National Institute of Technical Teachers Training and Research, Chandigarh, India

 | https://doi.org/10.1515/9783112218846-010

10.1 Introduction

Piezoelectric materials represent the most critical categories of electromechanically active, structure-actuated, and energy-transducing functional materials for next-generation sensing, actuation, energy harvesting, and adaptive structural systems [1, 2]. As one of the most widely utilized smart electronic material platforms, piezoelectric materials continue to undergo technological evolution, material optimization, and market-driven performance enhancements [3]. The global piezoelectric materials market, currently valued at approximately USD 35.6 billion, increasingly demands manufacturing strategies capable of overcoming four fundamental and persistent constraints of conventional fabrication technologies: (i) geometric and topological design limitations leading to >50% material waste, (ii) incompatibility with hybrid and multi-material integration protocols, (iii) insufficient miniaturization accuracy, typically constrained to ±25 μm in conventional micromachining and subtractive microfabrication, and (iv) economically prohibitive cost structures for low-volume, customized, and small-batch production [4–7].

The evolution of piezoelectric materials has progressed from early quartz crystal oscillators (1920s) to ferroelectric perovskite ceramics, including barium titanate ($BaTiO_3$) and lead zirconate titanate (PZT), which exhibit piezoelectric charge coefficients of $d_{33} \approx 190–650$ pC/N, electromechanical coupling factors (k_t) in the range of 0.45–0.60, and mechanical quality factors (Q_m) exceeding 100–500, depending on dopant chemistry and microstructural engineering [8–10]. Polymer-based systems (such as polyvinylidene fluoride (PVDF) and PVDF-TrFE (Trifluoroethylene)) and ceramic-polymer composite structures with enhanced processability, mechanical properties, and flexibility are among the advancements reported in recent studies [11–13].

Lithium niobate ($LiNbO_3$, $T_c = 1{,}210$ °C) and Lead Magnesium Niobate - Lead Titanate (PMN-PT) are examples of single-crystal piezoelectrics with remarkable functional performance. However, their employment in intricate or mechanically compliant device architectures is limited by their intrinsic brittleness, heightened fracture sensitivity, and limited adaptability to freeform geometries [14]. Long-term fatigue resistance, device longevity, and strain transfer efficiency are all greatly diminished by structural rigidity. By employing additive manufacturing (AM) to develop a geometry-based material architecture, these limitations can be greatly reduced. The associated constitutive equations govern the electromechanical behavior of piezoelectric materials:

$$S = s^e T + d^T E$$

$$D = dT + \varepsilon^T E$$

where S stands for the mechanical strain, T for the mechanical stress, E for the electric field, D for the electric displacement, s^2 for the elastic compliance under constant electric field, d for the piezoelectric coefficient matrix, and ε^T for the dielectric permittivity at constant stress.

The parameters of the three-dimensional (3D) printing process significantly impact the material's microstructure, layer orientation, porosity, and poling efficiency. These factors strongly influence key performance metrics, including dielectric loss (tan δ), mechanical quality factor (Q_m), electromechanical coupling (k_t), voltage coefficient (g_{33}), and piezoelectric charge coefficient (d_{33}). To address the limits of conventional methods, 3D printing has emerged as a cost-effective, scalable, and digitally controlled approach for fabricating complex lattices, Functionally Graded Material (FGMs), metamaterials, and porous piezoelectric structures [15–19]. It builds components layer by layer from digital designs. This allows for unparalleled geometric freedom, the possibility to combine numerous materials, decreased material waste, quick prototyping, and mass customization [20]. The global AM market was valued at USD 24.8 billion in 2024 and is expected to grow at a compound annual growth rate of 21.0%, reaching USD 57.1 billion by 2028. AM technologies have matured substantially since their commercialization in the 1980s. The technology's adaptability for manufacturing both structural and functional components is evidenced by its extensive utilization across diverse sectors, such as electronics, biomedical devices, aerospace systems, structural health monitoring, soft robotics, and food-health engineering [21–27].

As demand for advanced manufacturing grows, 3D printing technologies have shifted from photopolymer-based prototyping platforms to high-performance systems for processing functional materials. These systems are capable of processing high-viscosity inks, ceramic-loaded suspensions, nanoparticle-reinforced composites, and rheologically engineered piezoelectric pastes [28–30]. This innovation developed a fundamental framework for incorporating piezoelectric materials into AM processes, facilitating the development of application-specific, multifunctional, and high-resolution electromechanical device architectures, distinguished by customizable performance metrics and improved design flexibility [31]. The parameters of 3D printing, including layer thickness, printing speed, extrusion pressure, curing energy density, nozzle diameter, raster orientation, infill density (ID), sintering temperature, and poling electric field strength, significantly influence the grain morphology, porosity distribution, crystallographic texture, residual stress profiles, and interlayer bonding strength of 3D-printed materials [32–34]. The piezoelectric performance parameters are directly affected by the microstructural features, including:

d_{33} = piezoelectric charge coefficient,
k_t = electromechanical coupling factor,
Q_m = mechanical quality factor,
ε_r = relative permittivity, and
tan δ = dielectric loss.

D_{33} values can be improved by 20–40% by increasing printing resolution and decreasing interlayer void content. Domain alignment and coupling efficiency are enhanced by optimized sintering processes ($k_t > 0.55$). Over time, dielectric loss ($\tan \delta > 0.05$) might reduce a device's dependability and efficiency. This occurs when the material is not completely cured and is overly porous [35]. Piezoelectric devices are made using a variety of AM techniques. These include fused deposition modeling (FDM) for polymer-based piezoelectric composites, stereolithography (SLA) for high-resolution ceramic lattices, direct ink writing for viscoelastic piezoelectric inks and composite filaments, and aerosol jet printing for micro-scale patterned piezoelectric films [36–38]. Each method has benefits and drawbacks in terms of electromechanical performance, throughput rate, rheology control of the material, deposition accuracy, and feature resolution.

The development of next-generation smart structures, adaptive sensors, flexible actuators, and high-performance energy-harvesting systems can be revolutionized by combining piezoelectric materials science, electromechanical modeling, and 3D printing process optimization, according to a recent literature review. By connecting printing process settings to performance measures (d_{33}, k_t, Q_m, dielectric loss, and permittivity), the AM process encourages the usage of architected piezoelectric devices in industry. This allows us to estimate how well they will perform and adjust the design based on data.

10.2 Bibliometric analysis and research gap

This chapter includes a bibliographic study that evaluates the scientific advancement, research trends, and thematic development of piezoelectric materials combined with 3D printing. To generate a comprehensive bibliometric map, the three most relevant keywords, "parameters," "3D printing," and "piezoelectric," were selected as the primary search terms for this bibliometric research. By entering these three keywords into the SCOPUS database (www.scopus.com), a complete dataset of relevant papers was created.

Based on the most important keywords in the SCOPUS database, research articles, review papers, conference papers, book chapters, editorials, notes, and books were extracted. Data was downloaded in ".csv" or ".ris" format for processing and display. SCOPUS reports 20 years of article coverage on piezoelectric materials and 3D printing. A keyword co-occurrence analysis was performed using VOSviewer (version 1.6.19) to analyze research subject relationships. This software visualizes bibliometric networks and identifies keyword-based research clusters. The SCOPUS dataset initially yielded 25,929 unique terms, which were subsequently refined to identify the most relevant and meaningful keywords for this study. To ensure the inclusion of only significant terms, a minimum occurrence threshold of a term "25" was applied. After applying the threshold criteria, the number of terms was reduced to 193 relevant keywords. Subsequently, 60% of the most relevant terms suggested by VOSviewer (116 terms) were further evaluated. After removing duplicated, irrelevant, and overly generic terms, a final set of 80 key terms was selected for the bibliometric mapping and

Table 10.1: Bibliometric analysis for important terms for 3D printing and piezoelectric, with the number of occurrences and their relevance score (courtesy: SCOPUS database).

ID	Term	Occurrences	Relevance score
1	3D printing technology	94	0.5052
2	Accuracy	70	0.9742
3	Actuator	91	0.5583
4	Barium titanate	79	1.3249
5	Bone regeneration	29	4.6507
6	Cell	88	0.5208
7	Ceramic	85	1.0963
8	Comparison	35	0.9259
9	Component	109	0.6954
10	Composite	152	0.338
11	Density	95	0.3146
12	Deposition modeling	38	0.7674
13	Diameter	55	1.4532
14	Digital light processing	29	4.3165
15	Dimensional printing	30	0.3332
16	Direct ink writing	27	0.4526
17	DLP	27	4.2144
18	Droplet	55	3.9128
19	Electric field	48	0.758
20	Electrical property	30	0.6691
21	Electrode	59	0.3318
22	Energy	114	0.3592
23	Energy harvester	60	0.5263
24	Energy harvesting	93	0.4828
25	Factor	70	0.7833
26	FDM	44	0.8902
27	Film	105	0.3544
28	Flexibility	96	0.2707
29	Frequency	135	0.8758
30	Functionality	59	0.186
31	Graphene	28	0.6255
32	Impact	48	0.8044
33	Influence	87	0.4206
34	Ink	85	0.8142
35	Inkjet printing	44	2.1742
36	Integration	69	0.3143
37	Investigation	66	0.367
38	Lead zirconate titanate	38	0.8321
39	Low cost	56	0.3484
40	Manufacturing	221	0.2427
41	Measurement	71	0.5611
42	Mechanism	98	0.9712
43	Nanoparticle	62	0.8338
44	Numerical simulation	27	1.9375

Table 10.1 (continued)

ID	Term	Occurrences	Relevance score
45	Parameter	157	0.3952
46	PDMS	26	0.3952
47	Phase	68	0.8304
48	Piezoelectric ceramic	52	1.5187
49	Piezoelectric coefficient	80	0.9013
50	Piezoelectric composite	45	0.6383
51	Piezoelectric device	57	0.3291
52	Piezoelectric energy harvester	34	1.6864
53	Piezoelectric performance	44	1.1765
54	Piezoelectric property	131	0.6308
55	Piezoelectric response	47	1.0273
56	Piezoelectric transducer	42	1.9246
57	Piezoelectricity	59	0.876
58	Poling	47	1.2814
59	Poly	70	1.5799
60	Polyvinylidene fluoride	80	1.2929
61	Porosity	43	0.703
62	Printer	77	0.5388
63	Prototype	58	0.8306
64	PVDF	144	1.4388
65	PZT	73	0.6144
66	Research	134	0.2499
67	Scaffold	73	2.565
68	Signal	84	0.6197
69	Simulation	66	0.6814
70	Smart material	27	0.2985
71	Stereolithography	37	0.5672
72	Strategy	111	0.5527
73	Substrate	68	1.2924
74	Test	77	0.3854
75	Thickness	65	0.4567
76	Tissue engineering	45	1.6655
77	Transducer	75	0.9439
78	Variety	44	0.2982
79	Vinylidene fluoride	33	2.9797
80	Viscosity	59	0.745

cluster analysis. Table 10.1 shows the most important terms, their occurrence counts, and their relevance scores.

In the VOSviewer visualization, the size of each node corresponds to the number of times a term appears in the dataset. Larger nodes indicate higher keyword frequency. Key terms include "manufacturing" (221), "parameter" (157), "composite" (152), "PVDF" (144), "frequency" (135), and "piezoelectric property" (131), highlighting their importance in this field. The strength of the connection between the terms is illustrated by the links that exist between the nodes. These links are based on the fact that the terms are frequently found together in published documents. The "number of occurrences" in bibliometric mapping is a measure of the frequency with which a particular keyword is present in the dataset, which is indicative of the topic's prominence and the focus of research. The relevance score represents how distinctive or representative a keyword is within a particular cluster. A higher relevance score suggests that a keyword is highly specific to a certain research theme.

In comparison, a lower score indicates that the term is more general and commonly used across multiple clusters. Based on Table 10.1, the term "manufacturing" has the highest number of occurrences (231), followed by "composites" and "PVDF" with 152 and 144 occurrences, respectively. Whereas "bone regeneration" has a maximum 4.6507 relevance score, followed by "digital light processing," which has a 4.3165 relevance score. Based on the SCOPUS database and Table 10.1, a bibliometric relationship map has been developed and is shown in Figure 10.1.

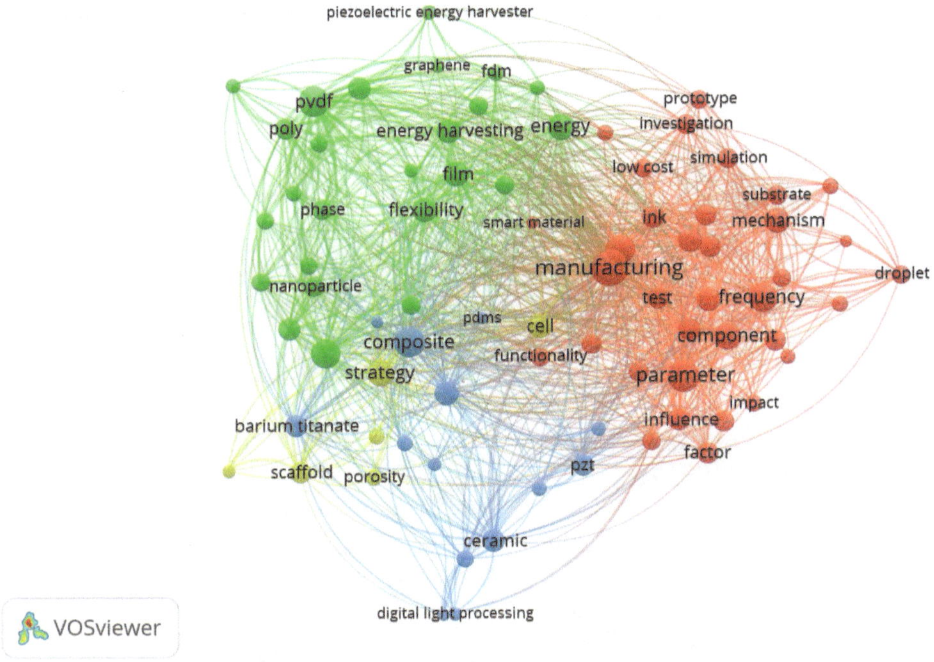

Figure 10.1: Bibliometric network relationship map of the most important terms related to this work (courtesy: SCOPUS database).

These 80 selected terms were organized into four distinct clusters, as shown in Figure 10.1, each representing a specific research theme within the broader field of piezoelectric 3D printing technologies. The clusters are distinguished, and the relationships between them are illustrated in the VOSviewer map using different colors (red, green, blue, and yellow). The research in cluster 1 (red) is predominantly focused on the optimization of process parameters, experimental investigations, and manufacturing techniques. This cluster highlights 35 key terms, including manufacturing, parameters, frequency, components, mechanisms, simulation, and testing, to optimize printing conditions and improve piezoelectric system performance. Cluster 2 (green) includes 25 keywords related to piezoelectric polymers and energy harvesting, including PVDF, graphene, nanoparticles, flexibility, and phase, and focuses on developing lightweight, flexible materials for wearable and low-power applications. Cluster 3 (blue), with 14 keywords, centers on structural aspects and ceramic-based piezoelectric materials, featuring terms such as density, barium titanate, composites, and piezoelectric properties. This group's research focuses on enhancing the electromechanical properties of ceramic materials and on investigating how to incorporate them into AM systems. Cluster 4 (yellow) represents biomedical and structural applications of piezoelectric materials. A total of six terms exist in this cluster. Terms such as cell, scaffold, porosity, and tissue engineering indicate the growing interest in using 3D-printed piezoelectric structures for biomedical implants, smart scaffolds, and regenerative medicine applications. Based on the SCOPUS database, a research gap analy-

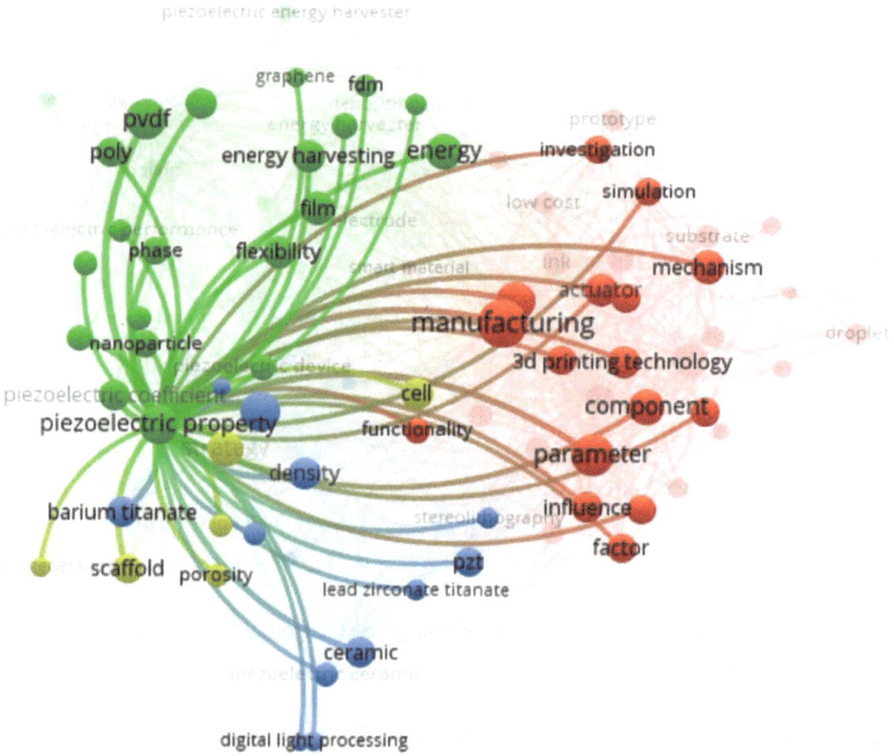

Figure 10.2: Research gap analysis using a bibliometric relationship network map diagram for "piezoelectric property" with other important terms (courtesy: SCOPUS database).

sis has been conducted for the keyword "piezoelectric property" alongside other most important keywords, as shown in Figures 10.1 and 10.2.

Based on the SCOPUS database, Table 10.1, and Figures 10.1 and 10.2, many terms related to piezoelectric properties that have been mostly studied in the past and published in the SCOPUS database are highlighted in a relationship graph. It has been observed that many words (parameter, 3D printing technology, component, ceramic, scaffold, porosity, phase, nanoparticle, PVDF, energy harvesting, etc.) are highlighted in Figure 10.2. It is also observed that lots of keywords are not directly connected with the keyword "piezoelectric property," which is not highlighted in Figure 10.3. Keywords that are not directly connected show that significantly less or no research and review work is published related to these terms on the SCOPUS database. Based on the SCOPUS database, Figure 10.3 has been drawn for the last 20 years of published articles related to this work.

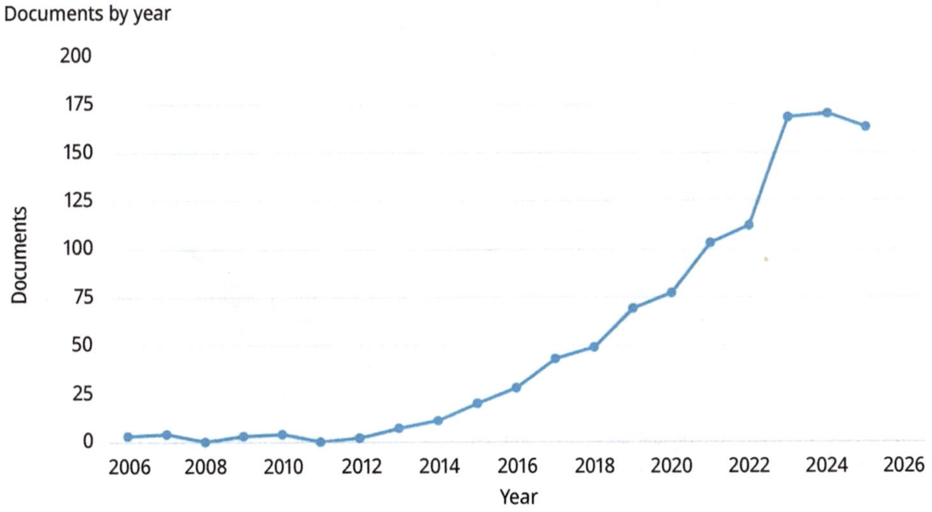

Figure 10.3: Last 20 years of published documents to this work (courtesy: SCOPUS database).

According to Figure 10.4, 66.5% of published documents are articles, 19.60% are conference papers, 8.40% are reviews, 2.5% are book chapters, 2.30% are conference reviews, and the remainders are editorial, note, book, letter, and so on. The examination of document kinds elucidates the characteristics of research contributions within this domain. The majority of publications are journal articles, accounting for approximately 66.5% of the total. This indicates that the majority of the research is published in peer-reviewed journals that focus on thorough experimental investigations, theoretical modeling, and advanced material characterization. Approximately 19.6% of all publications are conference papers. This shows that scholars are actively participating in international conferences to share early findings, new technologies, and new ideas.

About 8.4% of the papers are review papers. This shows a growing need to describe and critically analyze the rapidly developing body of research in this area. Also, a small portion of the entire dataset consists of book chapters (2.5%), conference reviews (2.3%), editorials, notes, books, errata, and retracted documents. According to the data, journal articles remain the most popular means of disseminating research findings on piezoelectric AM. Similar to Figure 10.4, Figure 10.5 was created using the SCOPUS database. A pie chart of document kinds by subject area is shown.

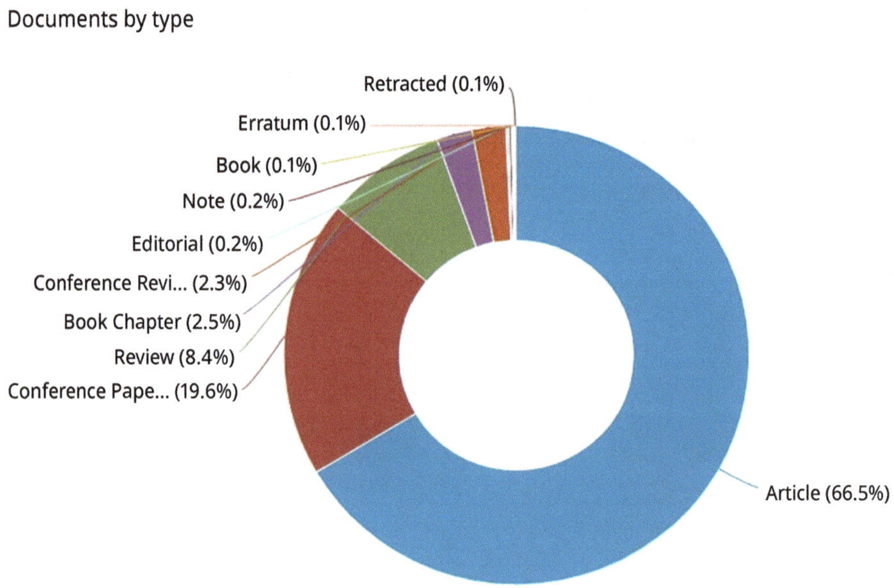

Figure 10.4: Pie chart for the last 20 years (2006–2025) of published documents by types related to this work (courtesy: SCOPUS database).

In total, 66.5% of published documents are articles, 19.60% are conference papers, 8.40% are review works, 2.5% are book chapters, 2.30% are conference reviews, and the other 10% are editorial, note, book, or letter types (Figure 10.4). The analysis of documents highlights the character of research contributions in this field. The majority of studies focus on advanced material characterization, theoretical modeling, and thorough experimental investigations, all published in peer-reviewed journals, as evidenced by the fact that journal papers account for about 66.5% of all publications. Conference papers account for about 19.6% of all publications, indicating that researchers regularly present initial findings, technical developments, and emerging concepts at international conferences.

The whole dataset consists of a smaller proportion of the following: book chapters (2.5%), conference reviews (2.3%), editorials, notes, books, errata, and retracted docu-

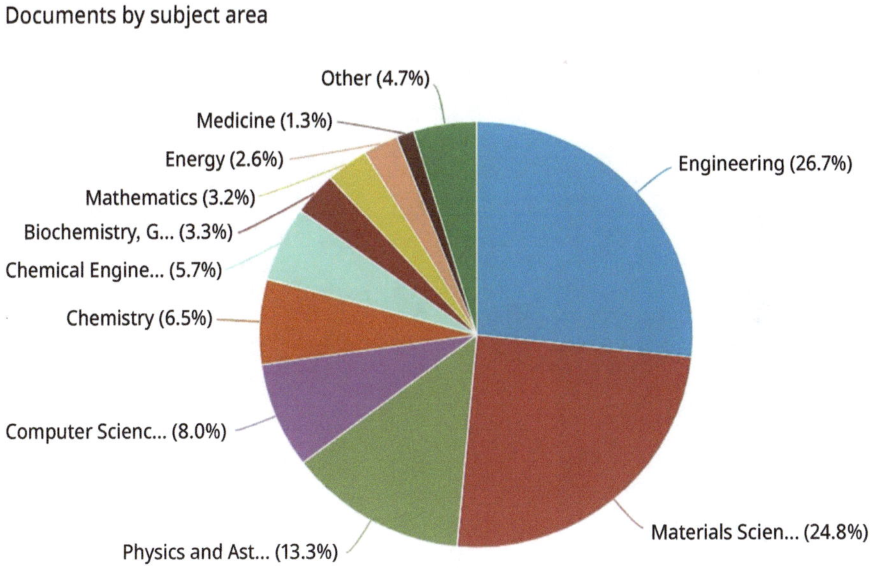

Figure 10.5: Pie chart for the last 20 years (2006–2025) of published documents according to their subject area related to this work (courtesy: SCOPUS database).

ments. These results imply that publishing research on piezoelectric AM in journals remains the most common way to share it. Using the SCOPUS database, Figure 10.5 creates a pie chart showing the subject areas of different types of papers, similar to Figure 10.4.

The subject-area distribution of publications indicates that the research domain is highly interdisciplinary, drawing on a variety of scientific and engineering disciplines. Figure 10.5 shows engineering as the largest field (26.7%), with emphasis on device design, manufacturing, and applications. Materials science (24.8%) follows, focusing on advanced piezoelectric materials like PVDF, PZT, and barium titanate. Physics and astronomy (13.3%) highlight fundamental studies of piezoelectric behavior. Computer science (8.0%), chemistry (6.5%), and chemical engineering (5.7%) support the synthesis and processing of materials. Emerging contributions from biochemistry, mathematics, energy, and medicine indicate growing applications in biomedical and energy devices. Figure 10.6 presents publication data for the top 10 countries from SCOPUS.

The USA has published 220 articles, followed by China with 351 and India with 77, as shown in Figure 10.6. China has the largest number of publications in this field. According to a bibliometric analysis, research on piezoelectric materials fabricated via 3D printing has increased substantially in recent years. The VOSviewer analysis also demonstrates strong connections among material development, AM processes, and application-oriented research. This identifies the most influential factors in AM technology that affect the fabricated piezoelectric properties of final products and

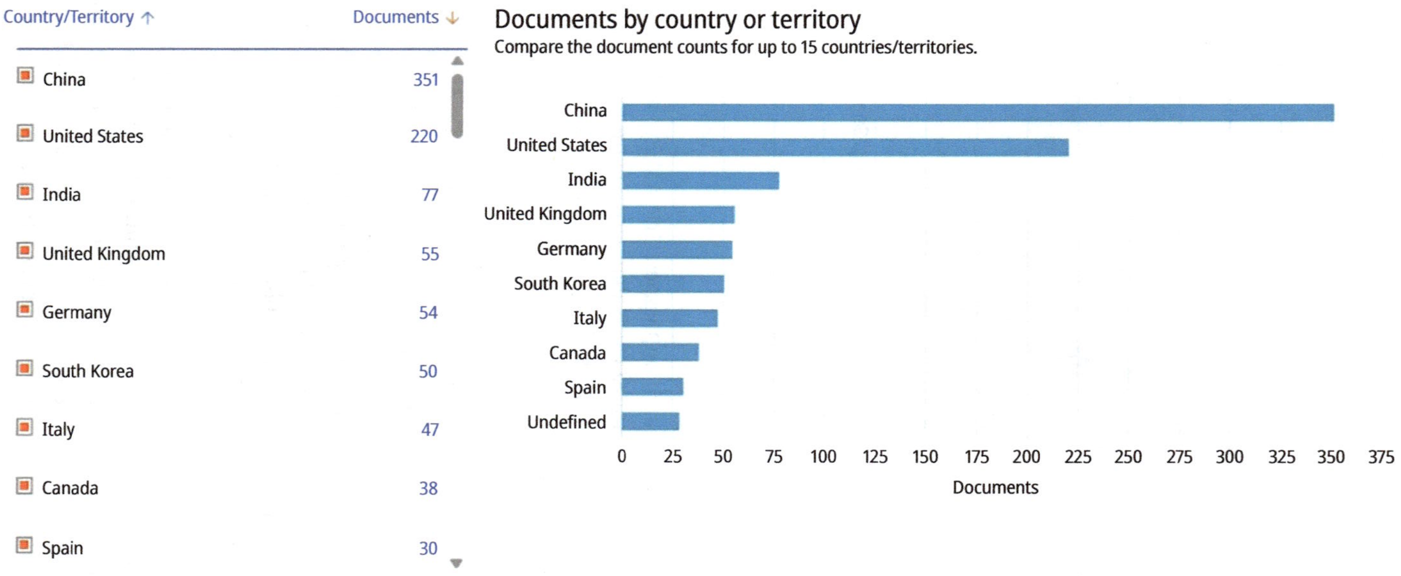

Figure 10.6: Bar diagram for the last 20 years (2006–2025) of the number of published documents by the top 10 countries (courtesy: SCOPUS database).

also outlines future opportunities in energy harvesting, biomedical implants, smart sensors, and flexible electronics.

10.3 Case study 1: 3D-printed piezoelectric materials for energy harvesting and sensing applications

AM has gained significant attention as an advanced processing technique for fabricating complex piezoelectric components with superior mechanical, thermal, and functional characteristics. Conventional/traditional fabrication techniques for piezoelectric materials, such as tape casting and bulk sintering, often limit geometric complexity and the integration of functional components. Three-dimensional printing enables the layer-by-layer fabrication of customized architectures with controlled microstructures and material compositions. This capability has facilitated the development of high-performance piezoelectric devices suitable for energy harvesting, sensing, and wearable electronics. Gong et al. [39] conducted extensive research and developed piezoelectric products via SLA-based 3D printing. This case study discusses the fabrication of barium titanate ($BaTiO_3$) ceramic piezoelectric structures using Liquid Crystal Display - LCD-based stereolithography (LCD-SLA). In this process, a printable slurry composed of approximately 70.1 wt% $BaTiO_3$ powder dispersed in a photocurable resin was prepared to achieve suitable rheological properties for printing. The green components produced by the SLA process were subsequently subjected to a multi-stage thermal treatment comprising debinding at 150–600 °C, followed by high-temperature sintering to obtain dense ceramic structures. After sintering, the printed samples demonstrated a relative permittivity (ε') of about 1965, a piezoelectric coefficient (d_{33}) of approximately 200 pC/N^{-1}, and a dielectric loss tangent of around 1.7%. These electrical characteristics are comparable to those obtained using conventional ceramic processing, indicating that SLA-based AM can produce high-quality piezoelectric ceramics with excellent electromechanical performance. Based on this case study, the fabrication process of piezoelectric materials using AM technology, their challenges, and different applications are shown in Figure 10.7.

Polymer-based piezoelectric materials have also been explored using AM due to their flexibility and compatibility with wearable devices. PVDF is one of the most widely used piezoelectric polymers because of its chemical stability and ability to form the polar β-phase responsible for piezoelectric behavior. In one study, PVDF-based composite resins were developed for SLA printing by dispersing PVDF particles within a photopolymer matrix. Mechanical testing under cyclic loading conditions revealed that the 2 wt% PVDF composite exhibited the highest piezoelectric coefficient ($d_{31} \approx 0.014$ pC/N), representing approximately a 46% improvement compared with the pure photopolymer resin. When the compressive force applied to the printed structure increased from 20 to 80 N, the generated piezoelectric current increased corre-

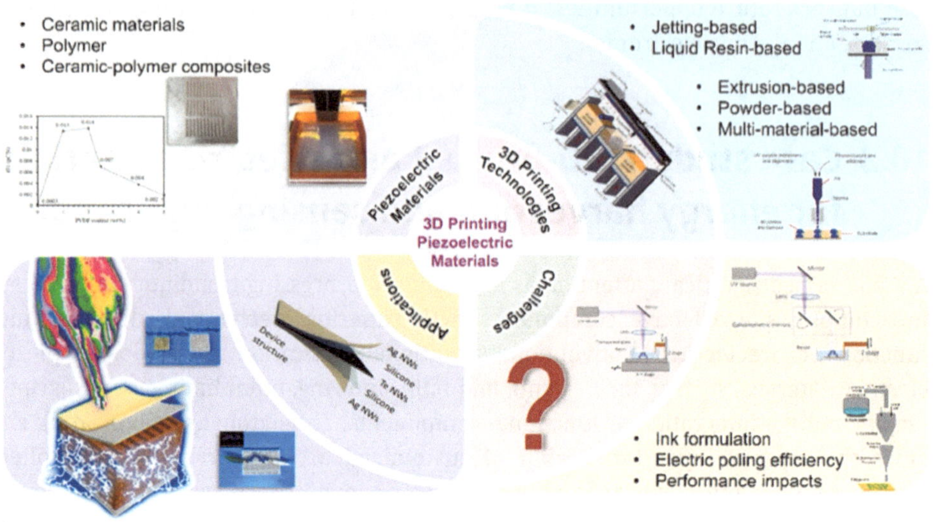

Figure 10.7: Schematic representation of materials, AM techniques, challenges, and applications of 3D-printed piezoelectric materials [39].

spondingly and reached approximately ±0.121 nA. Further increasing the load beyond 90 N resulted in mechanical failure of the printed specimen. When the polymer matrix was electrically poled at approximately 3 MV/m, the dipoles aligned, activating the piezoelectric response. The properties of piezoelectric materials depend a lot on the structure of their crystals and the makeup of the substance. PZT and other ceramic piezoelectric materials typically exhibit higher piezoelectric coefficients, reaching values of around 190 pC/N.

PVDF-based polymers usually have values between 20 and 30 pC/N. To improve the performance of polymer systems, researchers have developed PVDF-TrFE copolymers. The β-phase is stable in these copolymers, and they can reach piezoelectric coefficients of more than 100 pC/N. Piezoelectric values are higher for ceramics than for polymers, but polymers are better because they are lighter, more flexible, and can be made at low temperatures.

The fabrication of piezoelectric materials out of ceramic-polymer composites via inkjet printing: This method used $BaTiO_3$ nanoparticles in a PVDF matrix to improve dielectric properties and promote the formation of the β-phase. The printing ink contains 10 wt% $BaTiO_3$ nanoparticles. Adding roughly 65 g/L of dimethyl sulfoxide helped the nanoparticles spread and improved PVDF crystallization. The printed composite films contained approximately 78% β-phase content, as determined by Fourier transform infrared spectroscopy, thereby significantly enhancing the piezoelectric response. The printed films generated an output voltage of roughly 4 V when you tapped them with your finger. This work suggests that they could be applied to self-sustaining sensing and low-power energy harvesting.

Nanomaterials can now be used to create flexible piezoelectric generators using SLA and inkjet printing. Flexible conductive networks and active piezoelectric layers were created by combining silver and tellurium nanowires. The device produced electrical impulses that could be used for energy collection and sensing when mechanical tension was applied. When coupled to a 1-GΩ load resistance, the generator generated around 32.1 nW/m^2 of power density at about 20% mechanical strain. The gadget also maintained its electrical output steady after 1,000 s of cyclic mechanical loading, indicating that it is very durable and reliable for flexible electronic systems.

AM also enables the fabrication of advanced structural geometries, such as porous lattices, gradient structures, and multilayer piezoelectric architectures. These structural designs can enhance the stress distribution within the material, thereby improving the electromechanical conversion efficiency. Complex and difficult geometries/shapes fabricated using 3D printing enable optimization of mechanical compliance and electrical output, particularly for applications involving dynamic mechanical loading. High ceramic filler content may increase the viscosity of printing inks and cause nozzle clogging in jet-based printing techniques. Light scattering from ceramic particles can reduce the curing depth in photopolymerization-based printing processes. Powder-based methods can create porosity or microcracks during sintering, lowering dielectric and piezoelectric performance. Careful control of printing parameters, material composition, and post-processing steps such as annealing and poling is crucial for achieving better results.

This case study demonstrates that AM offers a flexible approach to producing complex piezoelectric devices with enhanced functionality. With optimized materials and design, 3D-printed piezoelectrics can deliver strong electromechanical performance, making them suitable for energy harvesting, biosensing, and structural health monitoring.

10.4 Case study 2: investigations of the 3D printing process parameters on the performance of the piezoelectric structures

AM increases the fabrication of complex piezoelectric structures using polymer composites such as PVDF with graphene (Gr) and barium titanate ($BaTiO_3$). This section outlines how process parameters affect mechanical, electrical, and piezoelectric performance [40]. PVDF, a semicrystalline electroactive polymer, is valued for its light weight, flexibility, and inherent piezoelectricity. Its behavior depends on the crystalline phase: the non-polar α-phase shows a weak response, whereas the polar β-phase, with aligned dipoles, exhibits strong activity. Studies report that ~90–95% β-phase significantly improves coefficients such as d_{33} and g_{33}. Mixing ceramic fillers like $BaTiO_3$, with high piezoelectricity (~1,700 pC/N), further enhances dielectric properties and polarization. Graphene contributes to mechanical strength and electrical conductivity.

Twin-screw extrusion (TSE) is used to blend the chemicals to form PVDF-based composites mechanically. To ensure nanofillers are uniformly dispersed throughout the polymer matrix, TSE is crucial. The barrel temperature (180–250 °C), screw speed (50–60 rpm), and feed content all significantly affect the rheological and mechanical characteristics of the extruded filament. Examples of optimal formulations with improved mechanical strength and electrical performance include PVDF (83–85 according to wt%), $BaTiO_3$ (15 according to wt%), and graphene (2 according to wt%) [41].

FDM fabricates products by joining layers layer by layer. Based on the literature review for this research work, Sharma et al. [41] concluded that the most important FDM settings for the fabrication of piezoelectric composite materials using PVDF thermoplastic polymers are infill speed (IS), ID, and infill angle (IA). Taguchi-based experimental design (L9 orthogonal array (3^3)) and Analysis of variance (ANOVA) show that the speed of the infill is responsible for 65.67% of the change in surface hardness. Lower ISs make layers stick together better and fill gaps, making the material stronger and spreading stress more evenly. ID also helps figure out how strong and compact the printed construction is. By increasing stiffness and decreasing porosity, a larger ID (nearing 100%) improves the piezoelectric response under mechanical loading. IA adds to anisotropic behavior even though it has less of an impact. Because load-bearing lines are more aligned, structures printed with 0°/90° raster orientations often have better mechanical qualities than those manufactured at ±45°.

Shore-D surface hardness, which in this study ranges from 65 to 77, is frequently used to assess the mechanical performance of 3D-printed composites, which can vary with processing conditions. At an IS of 50 mm/s, IA of 90°, and ID of 100%, the maximum hardness (≈76.89 Shore-D) is attained, indicating a more robust internal structure in these circumstances. Stress-induced polarization benefits from such increases in mechanical strength, thereby enhancing the piezoelectric response. The density, piezoelectric properties (d_{31}, g_{31}), and acoustic impedance value of different piezoelectric materials are shown in Table 10.2.

Table 10.2: Material properties of piezoelectric components.

Material	Density (g/cm^3)	d_{31} (pC/N)	g_{31} (Vm/N)	Acoustic impedance
PVDF	1.78	23	216	2.7
Nylon 11	1.15	0.26	2.5	2.9
PVC	1.45	0.7	6.6	3.27
PZT	7.5	1,200	10	30
$BaTiO_3$	5.7	1,700	5	30

Based on Table 10.2, it has been concluded that ceramic materials such as $BaTiO_3$ exhibit superior piezoelectric coefficients, but their brittleness limits their standalone

Table 10.3: Influence of FDM process parameters on surface hardness [41].

Experiment no.	Infill speed (IS) (mm/s)	Infill angle (°)	Infill density (ID) (%)	Hardness (Shore-D)
1	50	0	60	73.25
2	50	45	80	74.48
3	50	90	100	76.89
4	70	0	80	71.57
5	70	45	100	72.57
6	70	90	60	69.84
7	90	0	100	72.04
8	90	45	60	65.07
9	90	90	80	66.34

applications. PVDF-based composites provide an optimal balance between flexibility and functional performance. The design of experiments (Taguchi L9 (3_3) orthogonal array) for this case study is shown in Table 10.3, along with the Shore-D hardness.

According to Table 10.3, lower IS and higher ID together improve mechanical properties, which are the most significant for the performance of piezoelectric devices. Reinforcing graphene and $BaTiO_3$ in the PVDF matrix not only makes it stronger and better at handling heat, but also improves the dielectric and conductive pathways, facilitating charge creation and transport [42]. One can improve these reinforcements by adjusting the TSE parameters to distribute them evenly. Particle agglomeration can occur when dispersion is inadequate, leading to localized stress and reduced overall performance. Post-processing steps like electrical poling and thermal annealing are used to enhance piezoelectric performance. Poling aligns dipoles in the PVDF matrix, increasing polarization and improving piezoelectric coefficients. Even well-structured systems may underperform without proper post-treatment. The anisotropic nature of piezoelectric materials must be considered in device design, especially for sensors and energy harvesters, where directional response is crucial. A complex interaction between material composition, extrusion parameters, and FDM process variables affects the effectiveness of 3D-printed piezoelectric devices. By maximizing these parameters, it is feasible to create high-performance, versatile, and functional gadgets.

10.5 Conclusions

PVDF thermoplastic polymer is one of the most demanding piezoelectric materials for 3D printing applications due to PVDF uniquely bridging the gap between functional electronics and flexible structural materials, as compared to previously used different piezoelectric materials (like; ceramics), which are more brittle and very hard to provide to shape, whereas PVDF thermoplastic polymer provides better processability of plastic with the functionality of a sensor or energy harvester. Based on this work and previous reviews, several conclusions have been drawn, which are presented below:

- PVDF thermoplastic polymer is one of the most used materials that can be easily reinforced with other fillers. To enhance properties, many researchers and scientists have used high-performance ceramic nanoparticles (such as $BaTiO_3$ or PZT) to reinforce or mix them into the PVDF thermoplastic polymer to create a composite filament. PVDF is also a semicrystalline thermoplastic polymer that exists in multiple phases. In contrast, piezoelectric ceramics (e.g., PZT or $BaTiO_3$), which are brittle and prone to cracking, PVDF makes composite materials that are flexible, chemically resistant, and mechanically tough.
- According to Table 10.3, lower IS and higher ID together improve mechanical properties, which are the most significant for the performance of piezoelectric devices. Reinforcing graphene and $BaTiO_3$ in the PVDF matrix not only makes it stronger and better at handling heat, but also improves the dielectric and conductive pathways, making it easier to create and move charge.

References

[1] Bowen CR, Kim HA, Weaver PM, Dunn S. Piezoelectric and ferroelectric materials and structures for energy harvesting applications. Energy & Environmental Science. 2014;7(1):25–44.

[2] Safaei M, Sodano HA, Anton SR. A review of energy harvesting using piezoelectric materials: state-of-the-art a decade later (2008–2018). Smart materials and structures. 2019;28(11):113001.

[3] Shen D, Tao X, Koncar V, Wang J. A review of intelligent garment system for bioelectric monitoring during long-lasting intensive sports. IEEE Access. 2023;11:111358–77.

[4] Gharehpapagh B, Aliyev R, Gupta NK, Milek M, Zeidler H. Multi-material glass additive manufacturing: a review of materials, techniques, and integration challenges. Progress in Additive Manufacturing. 2026;1–21.

[5] Gotte M, Ramasreekanth PS. Advancements in Poly (Vinylidene Fluoride)-Based Piezoelectric Nanogenerators: Materials, Fabrication Techniques, and Applications. Polymer-Plastics Technology and Materials. 2026;65(1):60–99.

[6] Ligon SC, Liska R, Stampfl J, Gurr M, Mülhaupt R. Polymers for 3D printing and customized additive manufacturing. Chemical reviews. 2017;117(15):10212–90.

[7] Osanov M, Guest JK. Topology optimization for architected materials design. Annual Review of Materials Research. 2016;46(1):211–33.

[8] Berlincourt DA, Cmolik C, Jaffe H. Piezoelectric properties of polycrystalline lead titanate zirconate compositions. Proceedings of the IRE. 1960;48(2):220–29.

[9] Khorsand Zak A, Yazdi ST, Abrishami ME, Hashim AM. A review on piezoelectric ceramics and nanostructures: fundamentals and fabrications. Journal of the Australian Ceramic Society. 2024;60 (3):723–53.

[10] Panda PK, Sahoo B. PZT to lead free piezo ceramics: a review. Ferroelectrics. 2015;474(1):128–43.

[11] Kanwal A, Yang Y, Liu Z, Chen S, Wang F, Zhao X, Chew KH. A comprehensive review of piezoelectric BaTiO3-based polymer composites for smart tactile sensing. Emergent Materials. 2025;1–40.

[12] Nayak B, Jayasri S, Elorika P, Anwar S, Anwar S. Sustainable Smart Materials and Composite Structures. In Advanced Functional and Composite Materials 2025 (pp. 32–62). CRC Press.
[13] Wu H, Zhuo F, Qiao H, Kodumudi Venkataraman L, Zheng M, Wang S, Zhang Q. Polymer-/ceramic-based dielectric composites for energy storage and conversion. Energy & Environmental Materials. 2022;5(2):486–514.
[14] Yang Z, Zu J. Comparison of PZN-PT, PMN-PT single crystals, and PZT ceramic for vibration energy harvesting. Energy Conversion and Management. 2016;122:321–29.
[15] Alkunte S, Fidan I, Naikwadi V, Gudavasov S, Ali MA, Mahmudov M, Cheepu M. Advancements and challenges in additively manufactured functionally graded materials: a comprehensive review. Journal of Manufacturing and Materials Processing. 2024;8(1):23.
[16] De Marzo G, Fachechi L, Antonaci V, Mastronardi VM, Portaluri L, Todaro MT, De Vittorio M. On the measurement of piezoelectric d33 coefficient of soft thin films under weak mechanical loads: A rapid and affordable method. Materials & Design. 2024;247:113399.
[17] Dhakal N, Wang X, Espejo C, Morina A, Emami N. Impact of processing defects on microstructure, surface quality, and tribological performance in 3D printed polymers. journal of materials research and technology. 2023;23:1252–72.
[18] Fan J, Deneke N, Xu S, Newell B, Garcia J, Davis C, Nawrocki RA. Electric poling-assisted additive manufacturing technique for piezoelectric active poly (vinylidene fluoride) films: Towards fully three-dimensional printed functional materials. Additive Manufacturing. 2022;60:103248.
[19] Goulas A, Zhang S, Cadman DA, Järveläinen J, Mylläri V, Whittow WG, Engstrøm DS. The impact of 3D printing process parameters on the dielectric properties of high permittivity composites. Designs. 2019;3(4):50.
[20] Bănică CF, Sover A, Anghel DC. Printing the future layer by layer: a comprehensive exploration of additive manufacturing in the era of Industry 4.0. Applied Sciences. 2024;14(21):9919.
[21] Dhinesh SK, Robert JJ, Nair ST, Moni DS, Fowzeya SS, Kumar KS, Nagarajan P. Recent trends in additive manufacturing of electronic devices. Materials Today: Proceedings. 2022;66:928–41.
[22] Kasegn MM, Gebremedhn HM, Yaekob AT, Mesele E. The power of deoxyribonucleic acid and bio-robotics in creating a new global revolution: a review. Health Nanotechnology. 2025;1(1):3.
[23] Lu QY, Wong CH. Additive manufacturing process monitoring and control by non-destructive testing techniques: challenges and in-process monitoring. Virtual and physical prototyping. 2018;13 (2):39–48.
[24] Najmon JC, Raeisi S, Tovar A. Review of additive manufacturing technologies and applications in the aerospace industry. Additive manufacturing for the aerospace industry. 2019;7–31.
[25] Singh S, Ramakrishna S. Biomedical applications of additive manufacturing: Present and future. Current opinion in biomedical engineering. 2017;2:105–15.
[26] Song D, Chen X, Wang M, Wu Z, Xiao X. 3D-printed flexible sensors for food monitoring. Chemical Engineering Journal. 2023;474:146011.
[27] Tan HW, Tran T, Chua CK. A review of printed passive electronic components through fully additive manufacturing methods. Virtual and Physical Prototyping. 2016;11(4):271–88.
[28] Lasmi S, Bouchoul B, Marcos-Fernández ÁA, Lamiri L, Hamidouche M, Zoukrami F, Guerba H. Investigation on 3D-Printed Mesoporous Silica/Photocurable Resin-Based Nanocomposites: Mechanical, Morphological, and Thermal Properties. Journal of Materials Engineering and Performance. 2025;34(24):29827–38.
[29] Sacchi F, Colucci G, Bondioli F, Sangermano M, Messori M. bio-based photopolymers for additive manufacturing. Journal of Materials Science. 2025;60(27):11191–220.
[30] Wang Y, Ding Y, Yu K, Dong G. Innovative polymer-based composite materials in additive manufacturing: a review of methods, materials, and applications. Polymer Composites. 2024;45 (17):15389–420.

[31] Shen W, Zhang P, Li W, Qin H. Additive manufacturing for functional design: a review of capabilities, strategies and applications. Rapid Prototyping Journal. 2025;1–27.

[32] Koroglu L, Ayas E, Ay N. 3D printing of polyvinylidene fluoride-based piezoelectric nanocomposites: an overview. Macromolecular Materials and Engineering. 2021;306(10):2100277.

[33] Park J, Lee DG, Hur S, Baik JM, Kim HS, Song HC. A review on recent advances in piezoelectric ceramic 3D printing. In Actuators 2023 (Vol. 12, No. 4, p. 177). MDPI.

[34] Renteria A, Fontes H, Diaz JA, Regis JE, Chavez LA, Tseng TL, Lin Y. Optimization of 3D printing parameters for BaTiO3 piezoelectric ceramics through design of experiments. Materials Research Express. 2019;6(8):085706.

[35] Liu Y, Yuan S, Niu P, Zhang Z, Sun C, Lin Y, Chu X. Multi-dimensional optimization of slurry and synergistic suppression of interlayer cracking by layer thickness in DLP 3D printing silica-based ceramic cores. Ceramics International. 2025.

[36] Chen C, Wang X, Wang Y, Yang D, Yao F, Zhang W, Hu D. Additive manufacturing of piezoelectric materials. Advanced Functional Materials. 2020;30(52):2005141.

[37] Isakov D. Application of Piezoelectric Polymers in Additive Manufacturing. In Piezoelectric Polymers and Polymer Composites for Advanced Applications 2025 (pp. 179–204). Cham: Springer Nature Switzerland.

[38] Musa N, Dahiru M, Singh NB. 3D Printed Piezoelectric Energy Harvesters: Materials, 3D Printing Techniques, and Applications. Piezoelectric Materials. 2025;617–75.

[39] Gong G, You Y, Shen H, Laghaei M, Wang Y, Li Y. 3D printing piezoelectric materials: Innovations, challenges, and future perspectives. Materials Today Sustainability. 2025;101257.

[40] Sharma R, Singh R, Batish A. Investigations for barium titanate and graphene reinforced PVDF matrix for 4D applications. Encyclopedia of Renewable and Sustainable Materials. 2020;1:366–75.

[41] Sharma R, Singh R, Batish A. PVDF-graphene-BaTiO3 composite for 4D applications. In 4D Printing 2022a (pp. 103–19). Elsevier.

[42] Sharma R, Singh R, Batish A, Ranjan N. On Dynamic Mechanical Analysis of Mechanically Blended 3D Printed Functional Prototypes of PVDF-BaTiO3-Gr Composite.Encyclopedia of Materials. Plastics and Polymers. 2022b;1:318–26.

[43] Scopus database, www.scopus.com, retrieved on February 12th, 2026.

Ranvijay Kumar*, Rupinder Singh

Chapter 11
Application perspectives of the piezoelectric material-based 3D-printed functional structures

Abstract: Three-dimensional printing is known for its ability to manufacture customized structures for applications in the automotive, aerospace, biomedical, electronics, industrial, structural, and nonstructural sectors. The use of engineering materials adds value by enabling the fabrication of functional structures in such applications. Piezoelectric materials are engineered materials used in such applications. The 3D printing of piezoelectric materials with complex geometries and embedded functionality enables the manufacturing of customized designs. Three-dimensional-printed piezoelectric structures are widely reported for the development of sensors, actuators, sensor-embedded implantable devices, lightweight sensing components, energy harvesters, Internet of Things sensors, and various wearable devices. This chapter highlights the applications and perspectives of piezoelectric materials in 3D printing. This chapter also outlines the effects of parametric variation on the piezoelectric performance of the manufactured structures. In addition, a case study has been presented on the process capability analysis of 3D-printed piezoelectric materials for sensing applications.

Keywords: 3D printing, piezoelectric materials, sensors, functional structures, energy harvesting

11.1 Introduction

Piezoelectric materials generate an electric charge when mechanical forces are applied. Due to their extraordinary properties, piezoelectric materials are suitable for a variety of applications, including sensors, actuators, energy harvesting, and biomedical devices [1, 2]. In the past, the piezoelectric material requirements for these applications were met using conventional manufacturing techniques such as molding and other processes. Lead zirconate titanate (PZT), barium titanate ($BaTiO_3$), and polyvinylidene fluoride (PVDF) are recognized materials that possess good piezoelectricity [3, 4]. However, their processing by conventional methods, such as thin-film deposition, sintering,

***Corresponding author: Ranvijay Kumar**, Mechanical Engineering Department, Marwadi University Research Center, Marwadi University, Rajkot, Gujarat, India, e-mail: ranvijayk12@gmail.com
Rupinder Singh, National Institute of Technical Teachers Training and Research, Chandigarh, India

 | https://doi.org/10.1515/9783112218846-011

and casting, can pose significant problems. These problems concern the limitations of these methods for preparing complex geometries; there is significant material waste, and a challenge in minimizing size and design alterations. In recent years, 3D printing has emerged as one of the most important technologies for processing piezoelectric materials in their respective applications [5–7]. For the 3D printing of functional structures, PZT, $BaTiO_3$, and PVDF-based materials have been recognized as established materials with excellent piezoelectric properties. Along with their excellent piezoelectric properties, these materials are also recognized for their reprocessability and recyclability. The aid of 3D printing can boost the reprocessing of these materials into specific geometries tailored to the intended applications. Electronic and sensor fabrication is mostly reported using 3D printing on these materials. Despite the good piezoelectric and reprocessing aid of these materials by 3D printing [8, 9], several 3D printing techniques have been reported recently for the reprocessing and formation of functional structures from these materials. Fused deposition modeling (FDM), direct ink writing, stereolithography (SLA), and inkjet printing are recognized 3D printing techniques for processing. FDM is one of the most economical and most reported processes for the preparation of functional structures of piezoelectric materials. For example, Rodrigues et al. [10] reported modular modeling for FDM 3D printing of a piezodisk for metamaterial-based designs [10]. PVDF is one of the most suitable materials for FDM 3D printing and has been reported for applications such as energy storage and harvesting [11]. Also, Kim et al. [12] reported the FDM-based 3D printing of a piezoelectric PVDF film for pressure sensing [12]. Megdich et al. [13] reported energy harvesting from multimodal auxetic structures made from PVDF/multiwalled carbon nanotubes (MWCNTs) using FDM-based 3D printing [13]. The combination of the PVDF and molybdenum disulfide (MoS_2) is reported for preparation of conformal piezoelectric sensors [14]. A polylactic acid (PLA)-$BaTiO_3$ composite is reported for piezoelectric applications and is also 3D printable [15]. The FDM offers low-cost printing solutions and can be used to print polymer and composite materials. The reported studies have not been limited to the preparation of functional prototypes alone, but have also reported integrated components for real applications. Park et al. [16] have reported a review of the piezoelectric ceramic-based 3D printing processes [17]. Kim et al. [18] have reported a multimaterial combination of MWCNTs/$BaTiO_3$/PVDF-based nanocomposites for energy storage applications by GDM-based 3D printing [18]. Along with the low cost and ease of formation, there is a surface quality concern that limits the use of FDM-based 3D printing, especially in precision-based devices or models. SLA is also one of the established 3D printing processes that offers precise manufacturing and a high surface finish; however, it is limited to a few materials. Sotov et al. [19] have reported the LCD-SLA-based 3D printing of $BaTiO_3$ for the pressure sensing applications [19]. Pazourek et al. [20] have reported the PZT-photopolymer resin-based composites structures by SLA 3D printing for the perspective applications in energy storage/harvesting and smart devices [20]. The SLA-based 3D printing of piezoelectric materials has the potential to fit for the biomedical applications [21]. In this regard, a number of studies have been reported for the preparation of the functional structures of piezoelectric materials by SLA-based 3D printing [8, 16, 22–24]. Along with the FDM- and SLA-based 3D printing of piezoelectric materials, inkjet printing is also widely reported for piezoelectric materials [25–29].

Previous studies have reported the applicability of 3D printing processes for various applications, along with reprocessing and mathematical modeling to prepare suitable structures for the application. Despite the complex geometric designs it processes, 3D printing enables the application-oriented manufacturing of products. This chapter highlights the various applications of 3D printing with piezoelectric materials.

11.2 Design strategies for piezoelectric 3D-printed structures

The piezoelectric sensors, actuators, and energy harvesters employed in different applications convert electrical charge when a force is applied [30]. The design of the structures plays an important role in the output of the piezoelectricity when the structures are mechanically deformed [31]. Conventional manufacturing techniques are limited in producing the required geometry capable of efficient piezoelectricity. Three-dimensional printing offers a design-friendly manufacturing approach for the complex structures of piezoelastic materials, which are essential for piezoelastic models [8, 32].

The piezoelectric response of the structures is largely dependent upon the nature of the stress distribution, the distribution of the electric field, mechanical compliances, and connectivity arrangements [33, 34]. Three-dimensional printing can help control the above-mentioned parameters to improve the piezoelectric performance of the prepared structures. Along with the structural arrangement, the distribution of porosity and the cellular architecture of the piezoelectric structures can also affect their piezoelectric performance. The porosity factors, such as pore size, nature of porosity (open vs. closed), density of porosity, and lattice structures (cubic, tetragonal, etc.), are responsible for the piezoelectric response [35–37]. The FDM, SLA, and inkjet-based 3D printing can manufacture the parts by controlling these parameters. Usually, these improved designs result in a reduced elastic modulus due to better strain, which is associated with a good lattice pattern. Due to the improved strain, the increased piezoelectric performance can be achieved [33, 38]. The 3D printing process is also capable of manufacturing vertical piezoelectric pillars and laminar and interpenetrating structures, which improve piezoelectric response. Along with these parameters, anisotropic design, stress distribution, aspect ratio, functionally graded layers/structure, and multimaterial printing can play an important role in improving piezoelectric response, which can be facilitated by 3D printing [39–42].

11.3 Energy-harvesting applications

The 3D printing of piezoelectric materials is one of the most suitable processes for the fabrication of piezoelectric materials, not just due to the materials' good performance but also due to the ability to manufacture complex geometries (for efficient harvesting), customization of materials and designs, manufacturing of lightweight structures, and rapid manufacturing [43, 44]. The major techniques that cover energy-harvesting applications are materials extrusion, photopolymerization, and inkjet-based 3D printing processes [29, 45, 46]. These applications include the manufacturing of smart wearables, self-powered monitoring devices, and implantable devices/sensors by these 3D printing techniques [21]. In the structural fields, 3D printing of piezoelectric materials is used for monitoring structural health and for harvesting energy from vibrations. The 3D-printed sensors and energy harvesters in structural systems, such as bridges, buildings, aircraft components, and wind turbines, convert mechanical vibrations, air pressure, and forces into electrical signals for sensing and energy generation [6, 47]. The embedded energy harvesters, with 3D-printed tiles in the housing, can be used for energy harvesting. In road infrastructure applications, embedded road harvesters can generate electrical energy when pressure is applied. Also, energy recovery generation from the pedestrian road is possible if 3D-printed energy harvesters are applied with tiles. So, these 3D printing processes utilize piezoelectric materials in smart infrastructure and Internet of Things systems, specifically designed for energy harvesting. The energy harvesting is also possible by the aircraft and automobile systems using the vibrations and tire pressure [48, 49]. In addition, the 3D-printed piezoelectric devices can also be useful to harvest energy by manufacturing microelectromechanical systems (MEMS) devices and microsensors. Overall, the tailored energy harvesters by 3D printing support the lower materials waste, offer embedding, and flexibility. However, the challenges are also there, such as lower piezoelectricity due to defects, poling process optimization, and the mechanical strength of such systems.

11.4 Sensing and monitoring applications

In recent years, studies have reported on the 3D printing of piezoelectric materials for sensing and monitoring applications, such as health monitoring of buildings and large structures, biomedical applications, pressure sensing, acoustic monitoring, environmental parameter monitoring, and industrial process monitoring [50–53]. In structural applications, 3D printing of piezoelectric structures is used to detect and diagnose cracks and to monitor vibrations in large structures [53, 54]. It also detects the impact on the composite panels and the fatigue monitoring. These detections and monitoring could be performed using 3D-printed, embedded, and integrated conformal sensors. In biomedical applications, 3D printing of piezoelectric materials offers a useful tool for various diagnostic purposes. For example, blood pressure and pulse are usually measured with piezoelectric sensors [52]. Respiration rate and motion analysis could be other application perspectives, along with monitoring muscle activity. Three-dimensional printing helps manufacture these sensors and provides customization and flexibility to the designs [55]. In pressure-sensing applications, 3D printing of piezoelectric materials has been reported most often. The 3D printing of tactile and pressure sensors, whether integrated or separate, is an example. Pressure and tactile sensors are generally used as devices for tough detection, measuring grip force, collision detection, and sensing and analyzing deformation. Three-dimensional-printed piezoelectric sensors can be applied to acoustic applications, such as fault detection, bearing monitoring, and noise level measurement [56, 57]. Three-dimensional printing supports these applications by enabling the integration of sensing components and the manufacturing of sensor components for analysis. Several environmental parameters, such as wind speed, rainfall, water flow, and seismic parameters, can be measured using 3D-printed piezoelectric sensors.

11.5 Actuation and motion control systems

Actuation and motion controls are shape- and geometry-related entities that require special attention during manufacturing. Three-dimensional printing enables the customization of geometries optimized for actuation and motion control [55, 58]. The development of 3D-printed piezoelectric materials enables precise actuation in aerospace and microdynamic applications. Also, it is being applied in robotics applications. Various piezoelectric materials, such as ceramics, polymers, and composites, have been reported for the preparation of such structures by 3D printing for actuation and motion control [6, 16]. Ceramics such as PZT, polymers such as PVDF, and certain polymer composites have been widely reported for the preparation of actuation and motion control systems via materials extrusion, vat photopolymerization, inkjet printing, and aerosol jet printing [1, 6, 16]. The 3D printing process configures and controls the piezoelectric materials into the desired shape and size by specific anisotropy, integrated electrodes, multimaterial additive manufacturing, and functionally graded structures. The actuation mechanism works on the principle of 'converse piezoelectric effect," which tells that "when the voltage is applied, it causes a change in shape." In fact, these structures can be used for 4D printing if programmed and controlled properly. The 3D-printed stack actuator can offer fast response and precision, potentially suitable for micropositioning and adaptive optics applications [21, 59, 60]. The biomorph/unimorph, flexure, and lattice actuators are common examples of actuators that can be 3D-printed using piezoelectric materials. In motion control applications, 3D-printed piezoelectric materials can be used for open-loop, closed-loop, and self-sensing actuation. In open-loop motion control, the valve operates directly, with no feedback mechanism (generally for low-cost, low-time-critical actuation). However, in a closed-loop system, the integrated feedback mechanism provides good motion control. These structures can be used for the strain gauge, capacitive sensors, and optical sensors [33, 61, 62]. Also, the self-sensing actuators provide smart sensing that can be manufactured from smart materials. In addition to these applications, 3D printing of piezoelectric materials is used for precision monitoring, soft robotics (e.g., robotic fingers and bioinspired designs), aerospace, and biomedical applications.

11.6 Biomedical and healthcare applications

In biomedical and healthcare applications, the 3D printing of piezoelectric materials is now one of the most important considerations [47]. Measurements such as pulse, blood flow, respiration, and muscle movement can be obtained using 3D-printed sensors prepared from piezoelectric materials [52, 63, 64]. There are several utilities, such as pacemakers, biosensors, and many conformal sensors, that have been widely reported in previous studies [65–67]. There are several healthcare applications of 3D-printed piezoelectric materials, such as tissue engineering, drug delivery, tissue regeneration, wearable sensors and gadgets, implantable sensors, sensor-integrated implants for tissue repair and monitoring, and ultrasound components [47, 50, 52, 68]. In tissue engineering applications, the scaffolds prepared using piezoelectric materials can be useful for bone growth, tissue repair, and improved mineralization [69]. Piezoelectric materials are self-powered and support nerve regeneration. There are several wearable applications for 3D-printed piezoelectric materials, such as the preparation of smart clothing, sensor-integrated skin patches, and sensor-integrated prosthetics for monitoring. Real-time health checkups or health monitoring are possible with such utilities [25, 47, 50]. Also, the application pressure and strain sensors in the implantable form have added more value. In the monitoring of intracranial, intraocular, and cardiovascular pressure, 3D-printed piezoelectric-material-based sensors are helpful. The 3D-printed piezoelectric sensors are also used in ultrasound and imaging machines. The controlled release of medicines and drugs can also be achieved through 3D printing of piezoelectric materials. Tactile sensing and motion detection are key features of 3D-printed piezoelectric materials, which can efficiently control haptic feedback, gait analysis, and real-time adaptive control of prosthetics [50, 70–72]. Also, biomedical applications, such as smart wound healing, are a major utility of 3D-printed piezoelectric materials.

11.7 Case study for the development of 3D-printed piezoelectric structures for sensing applications

The sensing capabilities of the conformal sensors are highly dependent on the dielectric constant. In fact, the design and size of the conformal sensors are determined by the material's dielectric properties. In the case of 3D-printed piezoelectric sensors, dielectric properties are the key factors in sensor development. The work reported by Kumar et al. [73] has developed PVDF-based 3D-printed (gel + materials extrusion) prototypes for biomedical sensing applications [73]. The 3D-printed PVDF structures have shown a dielectric constant of 2.91, which is suitable for 3D printing sensing parts.

Along with this sensor parameter, the mechanical, thermal, and binding properties were also investigated. For the development of precise structures and sensing outcomes, the 3D-printed object should exhibit high control quality parameters. The process capability analysis of the 3D-printed structure ensures the manufacture of precise parts for such sensing applications.

Table 11.1: Dielectric constant of the 3D-printed PVDF structures.

Sample no.	Dielectric constant
1	2.91
2	2.85
3	2.97
4	2.88
5	2.94
6	2.90
7	2.95
8	2.88
9	2.93
10	2.95

In this study, a process capability analysis of the dielectric constant was performed to assess its suitability for precision manufacturing and sensing applications. In this study, similar prototypes were developed using the same parameters to investigate the dielectric constant as reported by a previous study [73]. Table 11.1 shows the dielectric constant values of 3D-printed PVDF structures, obtained from repeated experiments. The limit of maximum fluctuation was sex as ±5% of the dielectric constant. Given the target value of 2.91, the upper specification limit was set at 3.05 and the lower specification limit at 2.76 (Figure 11.1). The observations indicate that the process can produce accurate results with good process capability indicators. The overall standard deviation was observed to be 0.0386, indicating good control over data fluctuations. Also, C_{p} and C_{pk} values were observed at 1.21 and 1.14, respectively, both greater than 1, indicating good process capability. The data for the dielectric constant are controlled and follow a normal distribution, and no single observation was outside the control limits. Overall, the 3D printing of PVDF materials using gel + material extrusion is suitable for precision manufacturing and provides very good process control.

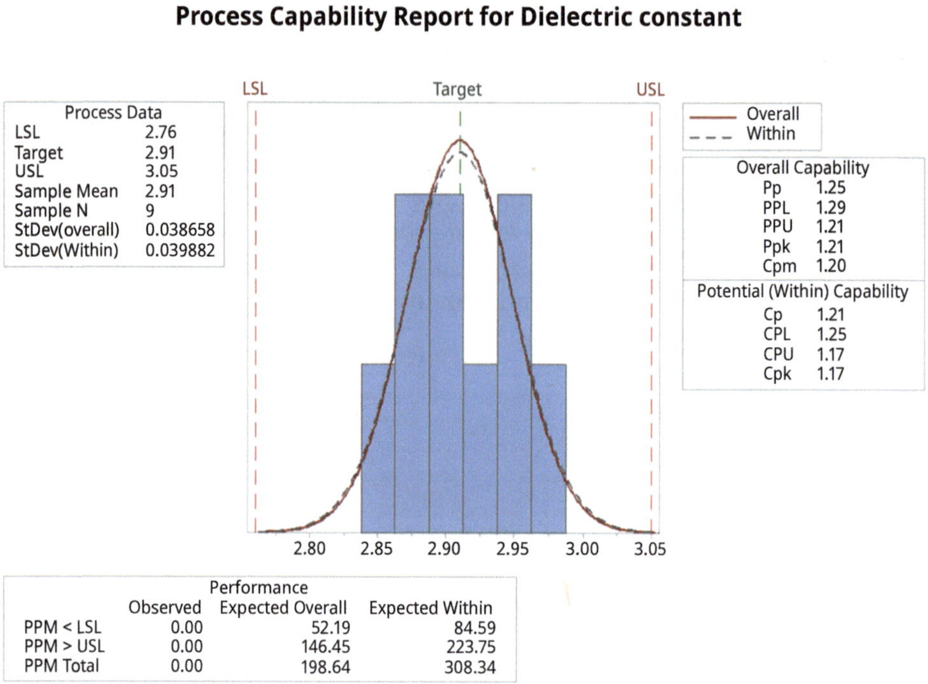

Figure 11.1: Process capability report for dielectric constant.

11.8 Conclusions

This chapter presents a state-of-the-art review of 3D printing of piezoelectric materials for the manufacturing of functional prototypes in energy-harvesting, biomedical, sensor, automotive, acoustics, and robotics applications. The following are the conclusions of this chapter.

Three-dimensional printing provides design freedom and enables the manufacturing of complex geometric shapes, which are responsible for the tuned piezoelectric behavior of 3D-printed parts and custom structures suitable for flexible and conformal sensors. Three-dimensional printing also results in minimal waste of piezoelectric materials compared to subtractive manufacturing.

Also, the ease of customization and ability to integrate provided by 3D printing result in improved performance. However, there are several challenges in developing 3D-printed functional structures of piezoelectric materials. Composite piezoelectric materials often exhibit nonuniform piezoelectricity due to several factors, thereby compromising sensing performance. Uniformity must be maintained to improve performance. Another challenge is the mass production of parts by 3D printing, control of material deposition patterns, and alignment control.

References

[1] Sekhar MC, Veena E, Kumar NS, Naidu KCB, Mallikarjuna A, Basha DB. A review on piezoelectric materials and their applications. Crystal Research and Technology. 2023;58(2):2200130.

[2] Uchino K. The development of piezoelectric materials and the new perspective. In Advanced piezoelectric materials 2017 (pp. 1–92). Elsevier.

[3] Brown C, Kell R, Taylor R, Thomas L. Piezoelectric materials, a review of progress. IRE Transactions on component parts. 1962;9(4):193–211.

[4] Dineva P, Gross D, Müller R, Rangelov T. Piezoelectric materials. In Dynamic Fracture of Piezoelectric Materials: Solution of Time-Harmonic Problems via BIEM 2014 (pp. 7–32). Springer.

[5] Bodkhe S, Ermanni P. Challenges in 3D printing of piezoelectric materials. Multifunctional Materials. 2019;2(2):022001.

[6] Cholleti ER (2018). A Review on 3D printing of piezoelectric materials. IOP conference series: materials science and engineering,

[7] Gong G, You Y, Shen H, Laghaei M, Wang Y, Li Y. 3D printing piezoelectric materials: Innovations, challenges, and future perspectives. Materials Today Sustainability. 2025;101257.

[8] Chen C, Wang X, Wang Y, Yang D, Yao F, Zhang W, Wang B, Sewvandi GA, Yang D, Hu D. Additive manufacturing of piezoelectric materials. Advanced Functional Materials. 2020;30(52):2005141.

[9] Sun D, Lu Y, Karaki T. Review of the applications of 3D printing technology in the field of piezoelectric ceramics. Resources Chemicals and Materials. 2023;2(2):128–42.

[10] Rodrigues GK, Silva MMD, Oliveira LPD. Modular modeling approach for FDM printed structures and piezo disks for metamaterial design. Latin American Journal of Solids and Structures. 2019;16 (07)e204.

[11] Koroglu L, Ayas E, Ay N. 3D printing of polyvinylidene fluoride based piezoelectric nanocomposites: an overview. Macromolecular Materials and Engineering. 2021;306(10):2100277.

[12] Kim H, Torres F, Wu Y, Villagran D, Lin Y, Tseng T-L. Integrated 3D printing and corona poling process of PVDF piezoelectric films for pressure sensor application. Smart Materials and Structures. 2017;26(8):085027.

[13] Megdich A, Habibi M, Laperrière L, Li Z, Abdin Y. Enhanced piezoelectric performance of PVDF/MWCNTs energy harvester through a 3D-printed multimodal auxetic structure for smart security systems. Materials Today Sustainability. 2024;27:100847.

[14] Islam MN, Rupom RH, Adhikari PR, Demchuk Z, Popov I, Sokolov AP, Wu HF, Advincula RC, Dahotre N, Jiang Y. Boosting piezoelectricity by 3D printing PVDF-MoS2 composite as a conformal and high-sensitivity piezoelectric sensor. Advanced Functional Materials. 2023;33(42):2302946.

[15] Murugan SS, Kattimani S. Effect of piezoelectric ceramic on natural frequency, structural, and thermal properties of additively manufactured PLA/BTO composite structure. Ceramics International. 2025.

[16] Park J, Lee D-G, Hur S, Baik JM, Kim HS, Song H-C. A review on recent advances in piezoelectric ceramic 3D printing. Actuators. 2023.
[17] Maidin S, Muhamad M, Pei E. Feasibility study of ultrasonic frequency application on fdm to improve parts surface finish. Jurnal Teknologi (Sciences & Engineering. 2015;77(32).
[18] Kim H, Johnson J, Chavez LA, Rosales CAG, Tseng T-LB, Lin Y. Enhanced dielectric properties of three phase dielectric MWCNTs/BaTiO3/PVDF nanocomposites for energy storage using fused deposition modeling 3D printing. Ceramics International. 2018;44(8):9037–44.
[19] Sotov A, Kantyukov A, Popovich A, Sufiiarov V. LCD-SLA 3D printing of BaTiO3 piezoelectric ceramics. Ceramics International. 2021;47(21):30358–66.
[20] Pazourek J, Pražák O, Procházka R, Straka L, Veselý P. Development of Piezoelectric Polymer Composites For Stereolithography. IEEE Transactions on Components, Packaging and Manufacturing Technology. 2025.
[21] Zeng Y, Jiang L, He Q, Wodnicki R, Yang Y, Chen Y, Zhou Q. Recent progress in 3D printing piezoelectric materials for biomedical applications. Journal of Physics D: Applied Physics. 2022;55 (1):013002.
[22] Musa N, Dahiru M, Singh N. 3D Printed Piezoelectric Energy Harvesters: Materials, 3D Printing Techniques, and Applications. Piezoelectric Materials. 2025;617–75.
[23] Smirnov A, Tikhonov A, Dubinin O, Chugunov S, Shishkovsky I. Piezoelectric properties of the 3D-printed lead-free ceramics. Ferroelectrics. 2022;601(1):179–84.
[24] Wang W, Sun J, Guo B, Chen X, Ananth KP, Bai J. Fabrication of piezoelectric nano-ceramics via stereolithography of low viscous and non-aqueous suspensions. Journal of the European Ceramic Society. 2020;40(3):682–88.
[25] Abdolmaleki H, Haugen AB, Merhi Y, Nygaard JV, Agarwala S. Inkjet-printed flexible piezoelectric sensor for self-powered biomedical monitoring. Materials Today Electronics. 2023;5:100056.
[26] Glinsek S, Song L, Kovacova V, Mahjoub MA, Godard N, Girod S, Biagi JL, Quintana R, Schleeh T, Guedra M. Inkjet-printed piezoelectric thin films for transparent haptics. Advanced Materials Technologies. 2022;7(8):2200147.
[27] Kim S, Choi JH, Sohn DK, Ko HS. The effect of ink supply pressure on piezoelectric inkjet. Micromachines. 2022;13(4):615.
[28] Shah MA, Lee D-G, Lee B-Y, Hur S. Classifications and applications of inkjet printing technology: a review. IEEE Access. 2021;9:140079–102.
[29] Thuau D, Kallitsis K, Dos Santos FD, Hadziioannou G. All inkjet-printed piezoelectric electronic devices: energy generators, sensors and actuators. Journal of Materials Chemistry C. 2017;5 (38):9963–66.
[30] Gautschi G. Piezoelectric sensors. In Piezoelectric Sensorics: Force Strain Pressure Acceleration and Acoustic Emission Sensors Materials and Amplifiers 2002 (pp. 73–91). Springer.
[31] Luck R, Agba EI. On the design of piezoelectric sensors and actuators. ISA transactions. 1998;37 (1):65–72.
[32] Srinivasaraghavan Govindarajan R, Ren Z, Melendez I, Boetcher SK, Madiyar F, Kim D. Polymer nanocomposite sensors with improved piezoelectric properties through additive manufacturing. Sensors. 2024;24(9):2694.
[33] Sirohi J, Chopra I. Fundamental understanding of piezoelectric strain sensors. Journal of intelligent material systems and structures. 2000;11(4):246–57.
[34] Xue F, Hu J, Wang SX, He J. Electric field sensor based on piezoelectric bending effect for wide range measurement. IEEE Transactions on Industrial Electronics. 2015;62(9):5730–37.
[35] Ju M, Dou Z, Li J-W, Qiu X, Shen B, Zhang D, Yao F-Z, Gong W, Wang K. Piezoelectric materials and sensors for structural health monitoring: fundamental aspects, current status, and future perspectives. Sensors. 2023;23(1):543.

[36] Rybyanets AN. Porous piezoceramics: theory, technology, and properties. IEEE transactions on ultrasonics, ferroelectrics, and frequency control. 2011;58(7):1492–507.

[37] Zhou X, Zhou K, Zhang D, Bowen C, Wang Q, Zhong J, Zhang Y. Perspective on porous piezoelectric ceramics to control internal stress. Nanoenergy Advances. 2022;2(4):269–90.

[38] Gullapalli H, Vemuru VS, Kumar A, Botello-Mendez A, Vajtai R, Terrones M, Nagarajaiah S, Ajayan PM. Flexible piezoelectric ZnO–paper nanocomposite strain sensor. small. 2010;6(15):1641–46.

[39] Jin Y, Xue SA, He Y. Flexible Pressure Sensors Enhanced by 3D-Printed Microstructures. Advanced Materials. 2025;37(27):2500076.

[40] Lee C-K. Piezoelectric laminates: theory and experiments for distributed sensors and actuators. In Intelligent Structural Systems 1992 (pp. 75–167). Springer.

[41] Pei H, Chen Y, Lv Q, Peng Z, Wang X, Chen N, Zhang H. A novel microwave assisted multi-material 3D printing strategy to architect lamellar piezoelectric generators for intelligent sensing. Composites Part B: Engineering. 2024;280:111529.

[42] Zhao H, Chen X, Xin C, Zhao F, Cheng S, Lei M, Wang C, Zhang J, Chen X, Tian H. High-sensitivity and self-powered flexible pressure sensor based on multi-scale structured piezoelectric composite. Chemical Engineering Journal. 2025;519:164787.

[43] Duque M, Murillo G. Low-cost manufacturing of monolithic resonant piezoelectric devices for energy harvesting using 3D printing. Nanomaterials. 2023;13(16):2334.

[44] Pei H, Xie Y, Xiong Y, Lv Q, Chen Y. A novel polarization-free 3D printing strategy for fabrication of poly (Vinylidene fluoride) based nanocomposite piezoelectric energy harvester. Composites Part B: Engineering. 2021;225:109312.

[45] Chen X, Ware HOT, Baker E, Chu W, Hu J, Sun C. The development of an all-polymer-based piezoelectric photocurable resin for additive manufacturing. Procedia CIRP. 2017;65:157–62.

[46] Hall SE, Regis JE, Renteria A, Chavez LA, Delfin L, Vargas S, Haberman MR, Espalin D, Wicker R, Lin Y. Paste extrusion 3D printing and characterization of lead zirconate titanate piezoelectric ceramics. Ceramics International. 2021;47(15):22042–48.

[47] Li S, Shan Y, Chen J, Chen X, Shi Z, Zhao L, He R, Li Y. 3D printing and biomedical applications of piezoelectric composites: A critical review. Advanced Materials Technologies. 2025;10(5):2401160.

[48] Bowen C, Arafa M. Energy harvesting technologies for tire pressure monitoring systems. Advanced Energy Materials. 2015;5(7):1401787.

[49] Eshghi AT, Lee S, Sadoughi MK, Hu C, Kim Y-C, Seo J-H. Design optimization under uncertainty and speed variability for a piezoelectric energy harvester powering a tire pressure monitoring sensor. Smart Materials and Structures. 2017;26(10):105037.

[50] Ali F, Koc M. 3D printed polymer piezoelectric materials: transforming healthcare through biomedical applications. Polymers. 2023;15(23):4470.

[51] Ma G, Li Y, Wang L, Zhang J, Li Z. Real-time quantification of fresh and hardened mechanical property for 3D printing material by intellectualization with piezoelectric transducers. Construction and Building Materials. 2020;241:117982.

[52] Mandal A, Morali A, Skorobogatiy M, Bodkhe S. 3D Printing of Polyvinylidene Fluoride-Based Piezoelectric Sensors for Noninvasive Continuous Blood Pressure Monitoring. Advanced Engineering Materials. 2024;26(1):2301292.

[53] Quah TKN, Vo TVK, Tay YWD, Tan MJ, Wong TN, Li KHH. Real time assessment of smart concrete inspection with piezoelectric sensors. Electronics. 2023;12(18):3762.

[54] Kumar V, Singh R, Ahuja IS. Online health monitoring of repaired non-structural cracks with innovative 3D printed strips in heritage buildings. Materials Letters. 2022;327:133033.

[55] Li J, Zhang Y, Yan M, Zhong C, Zhao L, Zhai D, Luo H, Yuan X, Zhang D. 3D printing of flexible piezoelectric composite with integrated sensing and actuation applications. Advanced Powder Materials. 2024;3(5):100226.

[56] Domingo-Roca R, Tiller B, Jackson JC, Windmill JFC. Bio-inspired 3D-printed piezoelectric device for acoustic frequency selection. Sensors and Actuators A: Physical. 2018;271:1–8.
[57] Zeng Y, Jiang L, Sun Y, Yang Y, Quan Y, Wei S, Lu G, Li R, Rong J, Chen Y. 3D-printing piezoelectric composite with honeycomb structure for ultrasonic devices. Micromachines. 2020;11(8):713.
[58] Ruiz-Díez V, García-Caraballo JL, Hernando-García J, Sánchez-Rojas JL. 3D-printed miniature robots with piezoelectric actuation for locomotion and steering maneuverability applications. Actuators. 2021.
[59] Baek S-W, Cho N-G, Lee D-H. Fabrication of a piezoelectrically driven micropositioning 3-DOF stage with elastic body using a multi-material 3D printer. Rapid Prototyping Journal. 2020;26(9):1579–91.
[60] Ferrara-Bello A, Vargas-Chable P, Vera-Dimas G, Vargas-Bernal R, Tecpoyotl-Torres M. XYZ micropositioning system based on compliance mechanisms fabricated by additive manufacturing. Actuators. 2021.
[61] Paulish AG, Zagubisalo PS, Barakov VN, Pavlov MA, Poyarkov AV. Piezo-optical transducer for high sensitive strain gauges. IEEE Sensors Journal. 2018;18(20):8318–28.
[62] Schitter G, Stark R, Stemmer A. Sensors for closed-loop piezo control: strain gauges versus optical sensors. Measurement Science and Technology. 2002;13(4):N47–N48.
[63] Jin L, Liu Z, Altintas M, Zheng Y, Liu Z, Yao S, Fan Y, Li Y. Wearable piezoelectric airflow transducers for human respiratory and metabolic monitoring. Acs Sensors. 2022;7(8):2281–92.
[64] Parupelli SK, Desai S. The 3D printing of nanocomposites for wearable biosensors: Recent advances, challenges, and prospects. Bioengineering. 2023;11(1):32.
[65] Dong L, Closson AB, Jin C, Nie Y, Cabe A, Escobedo D, Huang S, Trase I, Xu Z, Chen Z. Multifunctional pacemaker lead for cardiac energy harvesting and pressure sensing. Advanced healthcare materials. 2020;9(11):2000053.
[66] Kim J-H, Yoon H, Viswanath S, Dagdeviren C. Conformable piezoelectric devices and systems for advanced wearable and implantable biomedical applications. Annual Review of Biomedical Engineering. 2025;27(1):255–82.
[67] Kumar R, Kumar V, Husain M, Singh R, Kumar P. On Fabrication of Novel Polyvinylidene Fluoride BaTiO3-Cu-Doped ZnO-Based Filament for 3D Printing of Pacemakers. Journal of Materials Engineering and Performance. 2026;35(2):1477–90.
[68] Grinberg D, Siddique S, Le MQ, Liang R, Capsal JF, Cottinet PJ. 4D Printing based piezoelectric composite for medical applications. Journal of Polymer Science Part B: Polymer Physics. 2019;57 (2):109–15.
[69] Chen A, Su J, Li Y, Zhang H, Shi Y, Yan C, Lu J. 3D/4D printed bio-piezoelectric smart scaffolds for next-generation bone tissue engineering. International Journal of Extreme Manufacturing. 2023;5 (3):032007.
[70] Ali MA, Hu C, Yttri EA, Panat R. Recent advances in 3D printing of biomedical sensing devices. Advanced Functional Materials. 2022;32(9):2107671.
[71] Chang S-M, Hur S, Park J, Lee D-G, Shin J, Kim HS, Song SE, Baik JM, Kim M, Song H-C. Optimization of piezoelectric polymer composites and 3D printing parameters for flexible tactile sensors. Additive Manufacturing. 2023;67:103470.
[72] Liu C, Huang N, Xu F, Tong J, Chen Z, Gui X, Fu Y, Lao C. 3D printing technologies for flexible tactile sensors toward wearable electronics and electronic skin. Polymers. 2018;10(6):629.
[73] Kumar V, Singh G, Husain M, Kumar R, Meena H, Singh R, Hashmi M. Some investigations on hybrid 3D printed PVDF-based functional prototypes. Next Materials. 2025;8:100868.

Vinay Kumar*, Rupinder Singh

Chapter 12
Approach toward fully printed and integrated piezoelectric devices by 3D printing

Abstract: In the past two decades, novel studies have been reported on the fabrication of smart materials in which piezoelectric properties significantly contribute to the conversion of mechanical properties into electric signals to develop pressure sensors, tactile sensors, and condition health monitoring equipment for biomedical applications. But hitherto, very little has been reported on the fabrication of 3D-printed, fully functional, and integrated piezoelectric devices for engineering and biomedical applications. This chapter highlights the approach adopted for in-house fabrication of fully 3D-printed piezoelectric devices and integrated piezoelectric devices for biomedical applications. The results of investigations into rheological, thermal, mechanical, and morphological analysis of thermoplastic composites (such as polyethylene terephthalate, polyvinylidene fluoride, acrylonitrile butadiene styrene, thermoplastic polyurethane, and low-density polyethylene) have demonstrated good sensing capabilities. The tunable dielectric properties of these thermoplastic composites are highly acceptable for the manufacture of a variety of piezo-responsive smart sensors. The 3D-printed substrates for the proposed functionally graded composite matrices, prepared by 3D gel printing, laser-based 3D vat polymerization, 3D direct ink writing, extrusion-based fused filament fabrication, etc., may be used effectively for the design and development of conformal piezoelectric sensors. These fully 3D-printed and integrated piezoelectric sensors may be used for advanced online health/condition monitoring and digital twinning-based industrial and biomedical applications.

Keywords: Thermoplastic piezoelectric composites, in-house fabrication, recycling for good health, fully 3D-printed sensors, integrated composite sensors

12.1 Introduction

Investigations on solid materials (available as crystals and ceramic powders) that convert mechanical force applied to them into electrical output (voltage) have been

***Corresponding author: Vinay Kumar,** Department of Mechanical Engineering, Marwadi University Research Center, Marwadi University, Rajkot, Gujarat, India, e-mail: vinaykumar6192@gmail.com
Rupinder Singh, Department of Mechanical Engineering, National Institute of Technical Teachers Training and Research, Chandigarh, India

 | https://doi.org/10.1515/9783112218846-012

highlighted as novel piezoelectric materials for the preparation of various electronic devices [1]. The physical deformation of the surface has been regarded as a clear indication of shape and size change in the surface piezoelectricity-based piezoelectric thermoplastics and their composites for electronic signal transfer. Therefore, nanostructures of piezo-ceramics such as zinc oxide (ZnO), calcium carbonate ($CaCO_3$), and collagen have been identified as crucial elements for imparting the desired piezoresponsive properties in the base matrix for sensing applications [2, 3]. Investigations into the design and construction of transformers with multiple-response piezoelectric effects have shown that innovative energy harvesters can be fabricated with high efficiency and design conformality due to their ability to respond to a wide range of voltage gains at the resonance frequency (R_f) [4]. The "Scopus" database has indicated that very significant studies have been reported on physical applications of piezoelectric ceramics-based composites, whereas advanced manufacturing practices like 3D printing are least explored for the fabrication of an integrated and fully printed piezoelectric device. Based on this database, Figure 12.1 shows that 295 documents have been reported to date on the preparation of 3D-printed piezoelectric sensors for various electronic communication-based engineering and biomedical applications.

The experimental exploration of the theoretical physics underlying pressure- and tactile-sensitive piezoelectric sensors has shown that dielectric properties-assisted electrical response plays a significant role in the design of electroactive composites. For thin sensors, the pressure variation in the composite is less for a given time frame. Therefore, energy harvesters and transducers developed in this way are tested for optimized thicknesses over the range of applied pressure [5, 6]. The review literature reported on bio molecular sensing, bone regeneration, healthcare, human motion detection using nano-piezoelectric sensor devices prepared from various 3D printing techniques like selective laser sintering (SLS), stereolithography (SLA), fused deposition modeling (FDM), direct ink writing (DIW), ink-jet printing (IJP), digital light processing (DLP) have outlined that researchers have explored the potential of ceramics, semiconductors, amino acids, proteins, saccharides, amino polymers for acceptable crystalline behavior when integrated in the different composite materials to design and develop the proposed sensors [7, 8]. Based on this, Figure 12.2 shows the trend observed for documents reported on 3D-printed piezoelectric sensors in various publication sources such as *Nano Energy*, *Additive Manufacturing*, and *Sensors and Actuators A Physical*, in which such studies have been highly published in the past two decades.

The innovative design and development of piezoelectric materials are of great importance for the generation and detection of ultrasound waves. With such wave-detection techniques, novel defense equipment and medical instruments have been developed for the well-being of mankind [9]. The investigations have also focused on the development of lead-free (nontoxic) piezoelectric materials, biosensors, and implantable detector devices for human and veterinary surgical practices. The electromechanical analysis of electroactive piezoelectric materials, based on experiments

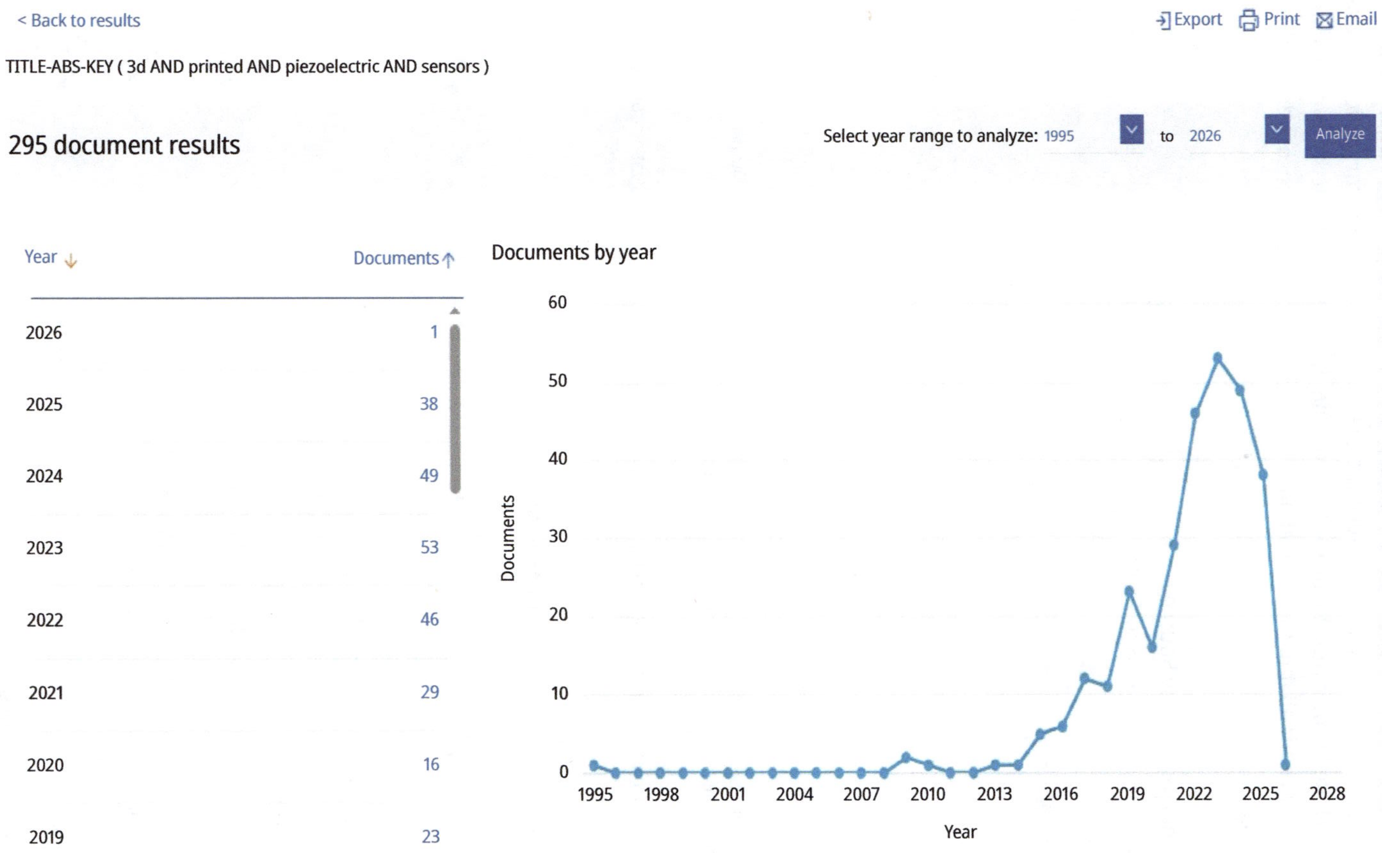

Figure 12.1: Trend obtained for the number of studies reported in the past two decades on 3D-printed piezoelectric sensors.

that investigated the movement of boundary elements, has been reported by some researchers as a novel approach to determine the acceptable ceramic for device fabrication [10, 11]. Regarding this, Figure 12.3 shows the documents reported by various authors that have highlighted the novel advantages of piezoelectric sensors enabled by 3D printing.

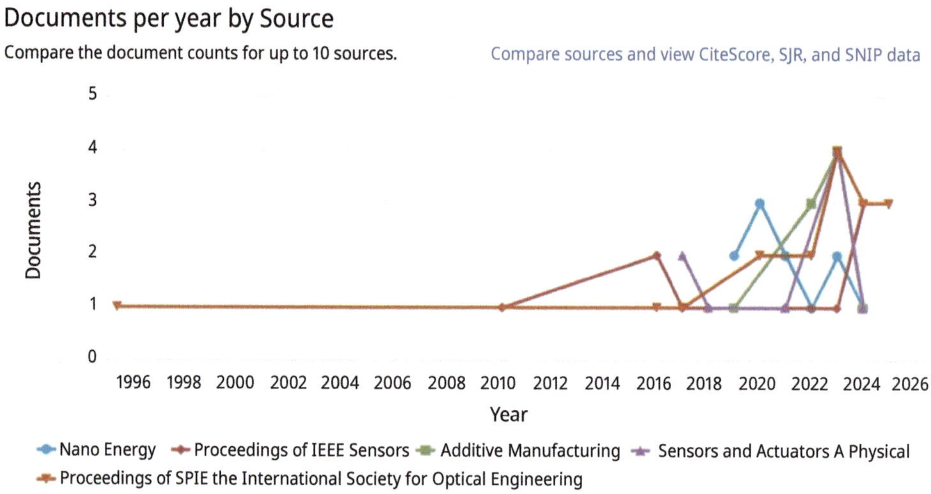

Figure 12.2: Documents reported on 3D-printed piezoelectric sensors in various publication sources.

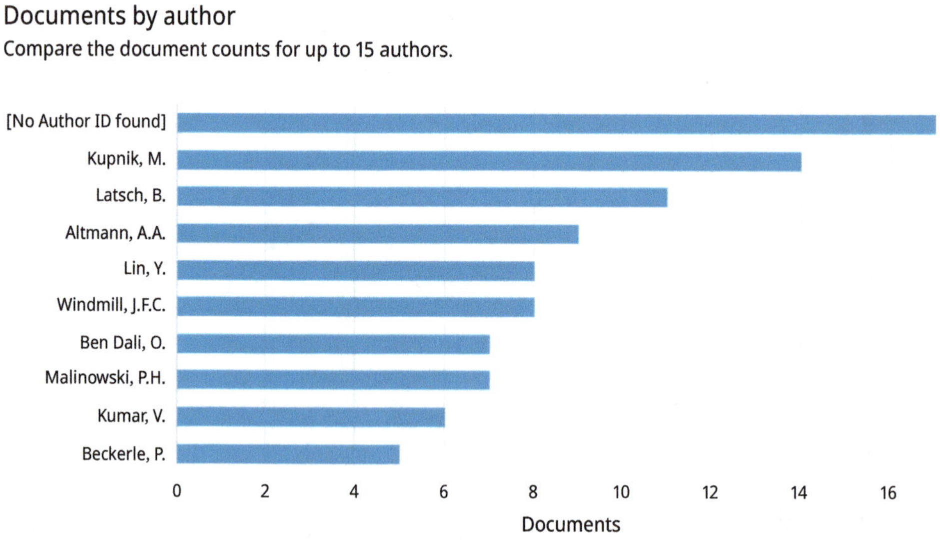

Figure 12.3: Number of documents reported on 3D-printed piezoelectric sensors by various authors.

Based on Figure 12.3, it may be stated that the experimental investigations performed on 3D printing technology for advanced piezoelectric energy-based engineering and biomedical applications, has paved a way for exploring the capabilities piezoelectric polymers and their composites (like polyvinylidene fluoride (PVDF), poly-L-lactic acid (PLLA), nylon-based polymers) as an alternative of standard ceramics for the fabrication of low piezoelectric coefficient-based 3D-printed sensors with programmability properties [12, 13]. 4D printing capabilities of conductive acrylonitrile butadiene styrene (ABS) and PVDF inks-based thermoplastic composite matrices also have been explored by some researchers for design and development of customizable polymer matrix-based smart piezoelectric solution with sensing properties for the repair and online health monitoring of heritage structures [14, 15]. Figure 12.4 presents a bar graph of the number of publications by various affiliations or institutions that are actively working in the area of advanced manufacturing systems assisted piezoelectric sensors.

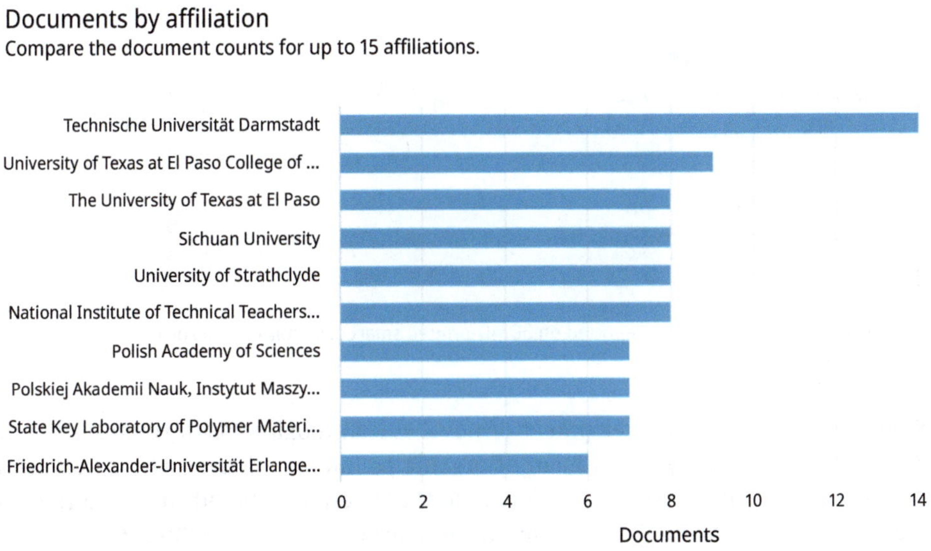

Figure 12.4: List of various affiliations associated with the development of advanced manufacturing systems assisted piezoelectric sensors.

It has been observed that the high thermal stability, dipole moment capacity, and semi-conducting properties of piezoelectric ceramics make them highly favorable for the preparation of sensors, actuators, piezoelectric nano power generators, energy harvesters for conformal electronic devices [16–18]. The studies reported on ZnO-reinforced PVDF matrix-based nano-fibers (tested for acceptable electrochemical properties, very high crystallinity of 74% and capability to achieve high β-phase) have been outlined as a novel biosensor for condition health monitoring of femoral artery and heart of veterinary animals, internal micropressure of patients, and flexible electronics used in fluid flow channels [19, 20].

The exploration of advanced applications of such smart materials indicates that researchers worldwide are actively investigating the possibilities of these smart transducers. Figure 12.5 shows the contributions of different countries to documents on 3D/4D-printed, smart, customizable piezoelectric sensors for a variety of engineering and biomedical applications.

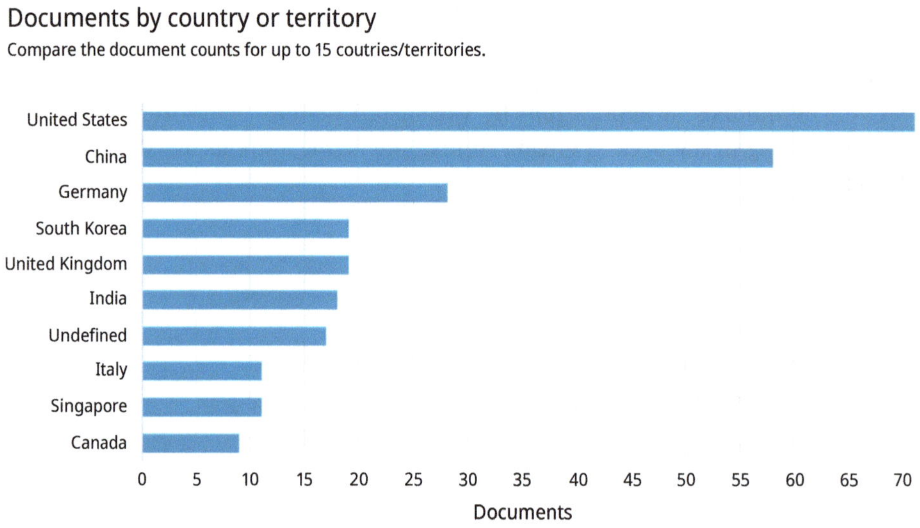

Figure 12.5: Number of studies reported on 3D/4D-printed smart customizable piezoelectric sensors by different countries.

Similar reports for the development of novel silicon dioxide (SiO_2) reinforced light-sensitive sensors (for the treatment of organic pollutants), thermosetting polymers, and electro-strictive lead-zirconate-titanate (PZT) ceramics-based fire sensors have outlined the advantages of such materials in smart civil structures [21–23]. In contrast to PZT-based ceramics, uniformly sintered barium-zirconate-titanate (BZT) – barium calcium titanate (BCT)-based ceramics have been explored for semiconductor fabrication-based on their acceptable dielectric, piezoelectric, and ferroelectric properties [24].

3D gel-printing-based fabrication of fully transparent piezoelectric sensors has been reported for fingerprint sensing and gesture monitoring. In this fully 3D-printed pressure sensor, all transparent layers act as transducers, with poly(methyl methacrylate) (PMMA) as the substrate. PVDF and polydimethylsiloxane (PDMS) contributed to imparting piezoelectric and elastic properties to the sensor, respectively [25]. This indicated that multiple material-based 3D-printed substrates may be explored with electrically conducting supporting fluids for the fabrication of freeform sensors and touch sensors used in electronic goods [26]. Fully 3D-printed piezoelectric sensors with acceptable electrical conductivity for advanced materials and manufacturing-assisted gel-based 3D printing, and FDM-based sensor fabrication, may be successfully utilized for dynamic tactile sensation and flexible electronics [27, 28]. Figure 12.6 presents the various types of such studies reported in book chapters, articles, review articles, and books on 3D-printed piezoelectric devices.

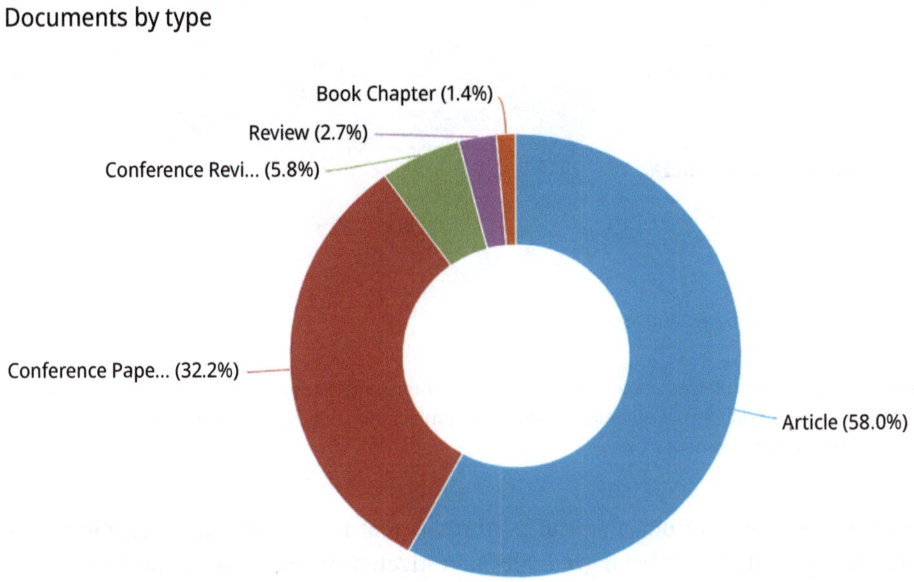

Figure 12.6: Pie chart distribution for the type of studies reported on 3D-printed piezoelectric sensors.

The investigations performed on design and fabrication of 3D-printed customizable and piezo-responsive-based programmable solutions for the repair and maintenance of heritage structures, structural health monitoring has indicated the challenges being faced by the conservators of various subject domains like civil engineering, material science, chemistry, additive manufacturing, electronics, and communication [29–31]. Based on this, Figure 12.7 highlights the contributions of various subject areas, which are highly aligned with studies on the design and preparation of different piezoelectric materials for 3D printing-based solution fabrication, which may be used effectively in smart sensor fabrication, digital twinning, and online condition monitoring.

Beyond fully 3D-printed one-way or two-way programmable sensors, researchers have explored integrated piezoelectric sensors, in which piezoelectric composites printed with a honeycomb structure were tested for ultrasonic devices and for wavy-substrate-based fiber arrays for signal transmission [32, 33]. The advantages of fully 3D-printed piezoelectric devices have been demonstrated in flexible and wearable

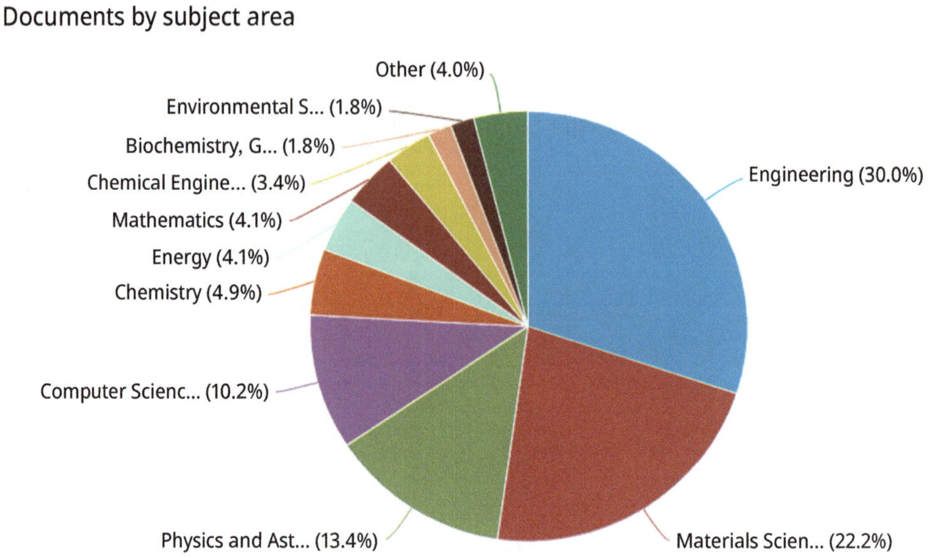

Figure 12.7: Pie chart-based presentations for the contribution of various subject domains like civil engineering, material science, chemistry, additive manufacturing, electronics, and communication in 3D-printed sensor development.

electronics. The electromechanical performance of fully 3D-printed piezoelectric sensors and actuators has been gradually investigated in recent years, and efforts have been made to develop more electroactive materials for such applications. The nanomaterials used in these types of actuators have been found crucial for inducing strain-based tuning of the optical properties of piezoelectric semiconductors [34]. The investigations performed on integrated 3D-printed piezoelectric sensor devices have outlined the focus laid on strain produced in the composite structure, resulting in high-performance actuator applications of such 3D-printed prototypes in maintenance of civil structures, condition monitoring of energy storage devices, and programming the piezoelectric actuators for high-performance electrical devices used in navigation systems [35]. The exploration of wide applications of fully 3D-printed and integrated piezoelectric sensors has highlighted the contributions of various international funding agencies, which have significantly advanced the research and development of these smart sensors. Figure 12.8 shows the contributions of different funding sponsors worldwide to the research and development of innovative piezoelectric sensors for engineering and biomedical sciences.

Figure 12.8: Bar graph-based contribution of different funding sponsors to the fabrication of fully integrated piezoelectric sensors.

12.2 Research gap and problem formulation

The literature review revealed that fully integrated and partially active piezoelectric sensors are widely reported for the preparation of 3D printing-based smart sensors, actuators, energy harvesters, and condition-monitoring systems. The database of 295 reported documents on 3D-printed piezoelectric sensors was analyzed in the VOS viewer to outline the various domains investigated by the researchers in developing these novel sensors. The processing of the publication database indicated that a total of 8,984 terms have been reported for 3D printing-based sensors. By including a minimum of 8 relevant terms (i.e., those related to highly relevant materials, their processing methods, and optimized mechanical, thermal, flexural, toughness, dielectric properties, and morphological characteristics), the analysis found that 199 terms meet the threshold value. By considering the 60% most relevant key research terms, a total of 120 key terms were identified, indicating the main research domains explored by the researchers for the proposed studies. Table 12.1 lists 120 reported keywords that are widely investigated for fully printed, 3D-printed piezoelectric devices.

The key terms outlined in Table 12.1 were grouped to form a web of keywords that indicate the core areas in which the proposed work has been pursued over the past three decades. Based on Table 12.1, Figure 12.9 shows the cloud of highly reported key terms on fully printed and integrated piezoelectric devices by 3D printing. This web of keywords highlighted that four core research areas, like material develop-

Table 12.1: List of key reported terms studied extensively for fully printed and integrated piezoelectric devices by 3D printing.

S. no.	Term	Occurrences	Relevance score
1	3D printing technology	15	0.4592
2	Amplitude	17	0.7501
3	Analysis of variance	58	0.3764
4	Array structures	25	0.3059
5	Barium titanate	23	1.1394
6	Batioinf3	8	1.1304
7	Batioinf3 INF	17	0.8598
8	BTO-CF	13	1.2487
9	Ceramic	19	0.576
10	Change in dimension	19	0.7681
11	Characterization	27	0.2985
12	Comparison	12	0.6286
13	Component	22	1.0149
14	Concept notes	15	0.5655
15	Conformality	13	5.7397
16	Control	20	1.4494
17	Cost	35	0.2598
18	Creation	9	1.2806
19	Customization	9	0.5428
20	Damage detection	14	2.9187
21	Data sets	33	1.0074
22	Density	32	0.4326
23	Detection	46	1.0065
24	Dimension	14	0.7911
25	Dinf33 INF	15	0.923
26	DLP	9	1.0246
27	Ease	9	0.4187
28	Electric field	12	0.7308
29	Emi	8	3.5485
30	Energy	39	0.4445
31	Energy harvester	29	0.6292
32	Energy harvesting	28	0.7318
33	Example	11	0.5922
34	Experiment	28	0.9705
35	Experimental result	12	0.4256
36	Extrusion	24	0.5631
37	Fdm	12	0.3699
38	Feasibility	17	0.4636
39	Filament fabrication	11	0.2771
40	Flexibility	35	0.7414
41	Form	12	0.5692
42	Fourier	9	1.1757
43	Frequency	45	0.6509
44	Ftir	11	1.1558

Table 12.1 (continued)

S. no.	Term	Occurrences	Relevance score
45	Functionality	19	0.3121
46	Graphene	11	0.339
47	Hand	13	0.4446
48	High sensitivity	19	0.5519
49	Impact	15	0.4919
50	Increase	19	0.2576
51	Inf	26	0.4988
52	Influence	17	0.3642
53	Ink	25	0.698
54	Internet	10	1.9119
55	Investigation	22	1.1334
56	Khz	15	1.9074
57	KPa	13	0.8059
58	Lead zirconate titanate	18	0.6274
59	Limitation	13	0.9475
60	Location	16	0.4161
61	Magnitude	11	0.3798
62	Matrix	28	0.372
63	Measurement	47	1.0294
64	Mechanical property	16	0.4019
65	MPa	10	0.8449
66	Object	13	2.1423
67	Polymer printing	81	0.6171
68	PCL	30	0.7326
69	PDMS	15	0.9954
70	PENG	8	2.5856
71	Phase	32	0.3861
72	Phase content	9	1.4521
73	Piezoelectric coefficient	30	0.6356
74	Piezoelectric nano-generator	13	2.2362
75	Piezoelectric performance	14	1.2806
76	Piezoelectric property	30	0.3814
77	Piezoelectric response	17	0.7135
78	Piezoelectric transducer	15	1.2598
79	Piezoelectricity	20	0.9202
80	Polylactic acid (PLA)	14	0.8898
81	Plate coating	13	2.4822
82	Point of care	12	0.8674
83	Poling	18	0.931
84	Polyurethane	21	1.3319
85	Polydimethylsiloxane	15	0.9954
86	Polylactic acid-composite	11	0.8658
87	Possibility	15	0.3464
88	Potential application	9	0.6343
89	Pressure sensor	23	0.4818

Table 12.1 (continued)

S. no.	Term	Occurrences	Relevance score
90	Print speed	11	0.4542
91	Printing process	13	0.4516
92	Printing technique	16	0.6292
93	Proceeding	17	5.8246
94	PVDF	55	0.5389
95	PVDF TRFE	12	0.7809
96	PZT	36	0.5017
97	Recent year	9	0.3494
98	Reliability	10	1.5701
99	Review	11	0.8204
100	Self	41	0.6864
101	Sensing application	13	0.686
102	Signal	44	0.6521
103	Smart material	11	0.4353
104	State	19	1.0008
105	Structural health monitoring	15	1.176
106	Tactile sensor	14	0.5303
107	Testing	12	1.3873
108	Internet of things	8	2.2024
109	Thermoplastic composites	17	5.8246
110	Transducer	25	0.5047
111	Type of sensors	26	0.3392
112	Vibration sample magnetometry	22	0.7896
113	Vinylidene fluoride	9	1.9077
114	PVDF-co-trifluoroethylene	8	2.088
115	Viscosity	14	0.4751
116	Volumetric analysis	9	1.4103
117	Voltage-current output	43	0.4183
118	Water absorption	9	0.6371
119	Wearable device	14	0.9369
120	Wearable electronic	14	1.1343

ment, sensor design, component characterization, and application areas, have been extensively explored by the investigators for the proposed 3D-, 4D-, and 5D-printed smart sensors. Some researchers have highlighted the novel biomedical scaffolding applications of a PLA-based thermoplastic composite matrix, along with acceptable integrated sensing properties for condition monitoring of treated patients [36].

Based on Figure 12.9, further insight into web keywords indicated that thermoplastic blends like polyvinyl alcohol (PVC), polypropylene (PP), and piezoelectric thermoplastic composite matrices of PVDF-calcium carbonate, PVDF-graphene, etc., have been investigated for 4D printing-based online health monitoring capabilities [37, 38]. Regarding such studies, Figure 12.10 shows studies on the preparation of different piezoelectric PVDF composite matrices using reinforcements such as barium titanate

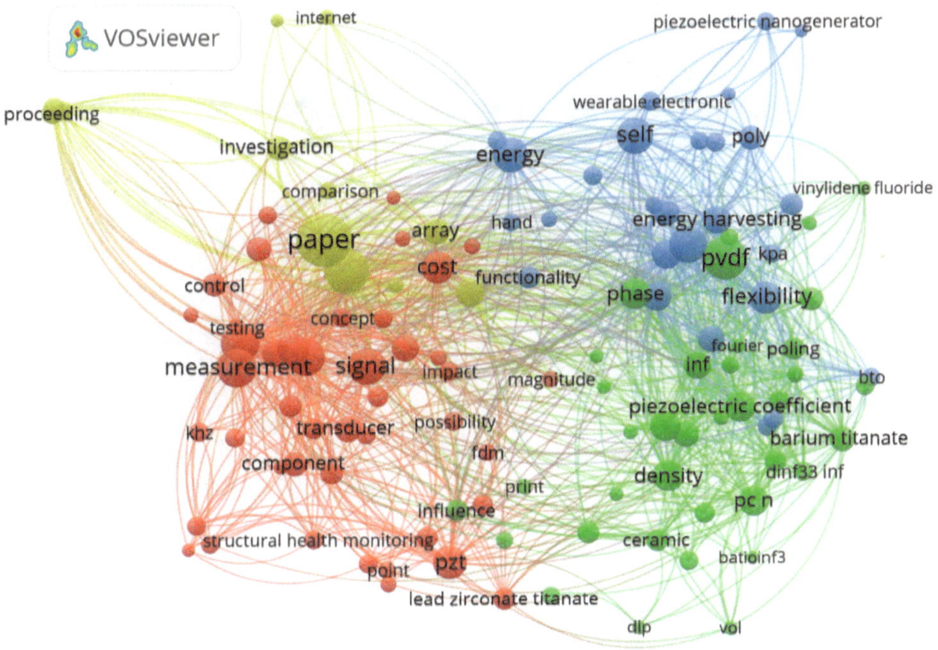

Figure 12.9: Cloud of highly reported keywords on designing, characterization, and fabrication of fully printed and integrated piezoelectric devices.

($BaTiO_3$) and ionic ceramics such as zinc oxide (ZnO) and Mn-doped ZnO for advanced engineering and biomedical applications.

Similarly, the overview of the 3D printing technology node in the web of nodes, shown in Figure 12.11, highlighted that partially and fully integrated 3D-printed sensors have been widely reported for wearable and flexible electronics. These energy-harvesting devices may be tuned to variable radio frequencies in the Bluetooth range, wireless fidelity (Wi-Fi), or satellite communication. The preparation of 3D/4D-printed PVDF-based smart energy storage devices and recycled polyethylene terephthalate (PET)-based low-cost food packaging sensors has shown great potential for 3D-printed piezo-responsive sensors in many commercial applications [39, 40]. But hitherto, very little has been reported on the fabrication of 3D-printed, fully functional, and integrated piezoelectric devices for engineering and biomedical applications. This chapter highlights the approach adopted for in-house fabrication of fully 3D-printed piezoelectric devices and integrated piezoelectric devices for biomedical applications.

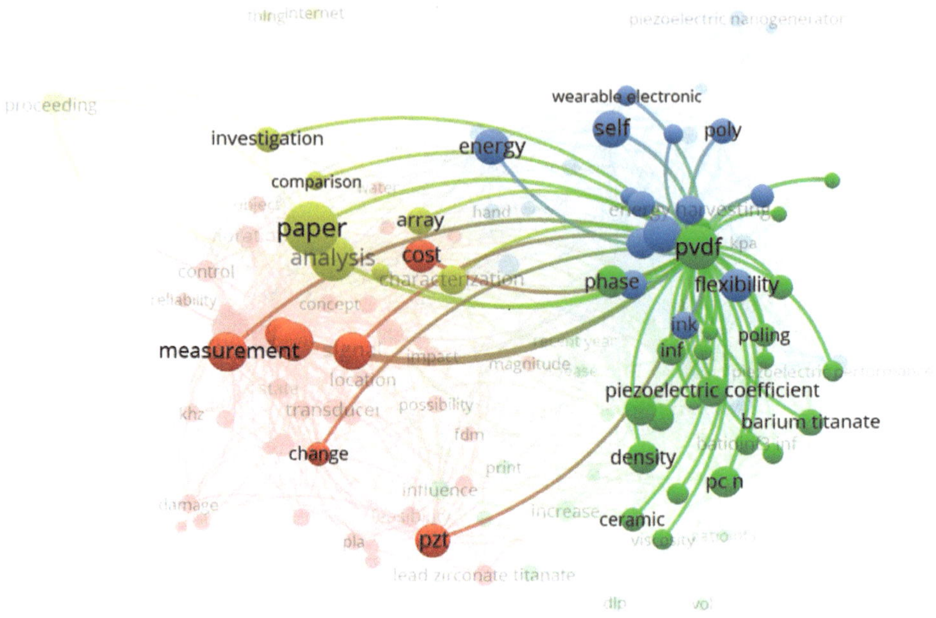

Figure 12.10: Studies associated with a web of keywords related to the preparation of piezoelectric PVDF composite matrices using different reinforcements.

12.3 Fully 3D-printed and integrated piezoelectric sensor fabrication

The approach to the fabrication of a fully 3D-printed piezoelectric sensor is based on the in-house preparation of a raw material with good dielectric properties for signal transmission. The process flow chart outlining the work methodology for the design and fabrication of fully 3D-printed piezoelectric sensors is shown in Figure 12.12. The work methodology outlined four main stages: composite material preparation, dimensional design of the sensor, 3D printing of a functional prototype, and characterization of the fabricated sensor for a suitable piezoelectric-based application. The thermoplastic matrix is first processed with piezoelectric ceramics to impart a dipole moment in the composite. Further, a direct current poling setup (DC poling unit) is used to determine the acceptable piezoelectric coefficient ($D33$) in the composite. The polarized surface of the piezo-responsive composite element is further tested for dielectric constant (ε_r) and loss tangent (tanδ) properties for sensor design. The composite prepared is 3D-printed to obtain the functional prototype. In the last stage, the sensor prototype is characterized for a suitable frequency range (e.g., Bluetooth (2.45 GHz), Wi-Fi (5 GHz), and satellite communication (beyond 5 GHz)).

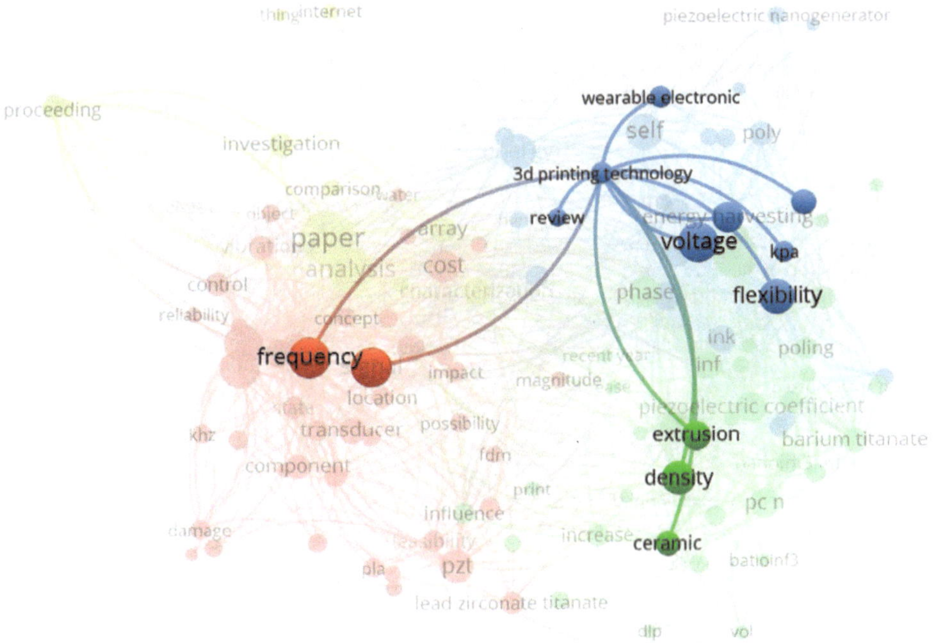

Figure 12.11: Studies associated with partially and fully integrated 3D-printed sensors.

The high-temperature-resistant, cost-effective 3D-printed sensors prepared from a ceramic-reinforced polymer composite matrix have been observed to be very useful for flexible and wearable electronic devices. As a result, conformal sensors may be fabricated for energy-harvesting systems, structural condition-monitoring devices, and implantable health-monitoring sensors for biomedical and surgical practices. Gel 3D printing and FDM-based fully 3D printing of PVDF, PLA, and their composite structures (embedded in polyethylene film) have been used for pressure sensing, optical sensors, and tactile (touch-based) actuators. The voltage (*V*), current (*I*), and resistance (*R*)-based *V–I* and *V–R* observations of fully 3D-printed tactile sensors have indicated very good accuracy for biomedical applications. For energy-harvesting 3D printing, MnO_2, $ZnCl_2$, graphite, and NH_4Cl salts have been used to reinforce PVDF composite matrices. The multimaterial 3D printing of these compositions in a dual-nozzle setup has been proposed as a novel approach to fabricate a piezo-responsive smart energy storage device with structural health monitoring capabilities [22, 39].

The approach of fully 3D-printed piezoelectric sensors, which leverages the full integration of multimaterial-based sensor fabrication, is a novel way to manufacture the entire sensor in a single run. This time-saving, cost-effective process enables engineers to design complex sensor geometries alongside the feature to prepare conformal sensors. By this approach, more customized, parametrically optimized 3D-printed sensors may be fabricated for both human and veterinary animals. The exploration

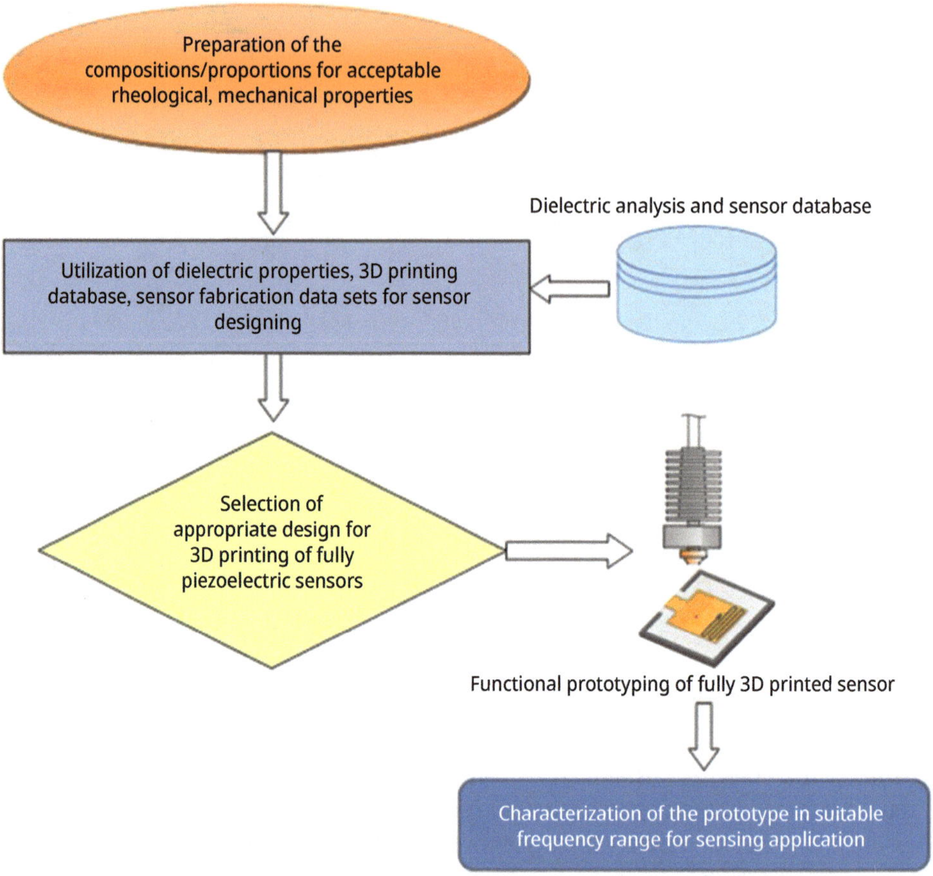

Figure 12.12: Process flow chart for the fabrication of fully 3D-printed piezoelectric sensors.

of hydroxyapatite (HAp)- and chitosan-reinforced PVDF and PLA composite sensors for the treatment of bovines and canines is an example of such innovative practices [41, 42].

In the case of integrated 3D printing of piezoelectric sensors: PP, PET, PVDF, ABS, thermoplastic polyurethane (TPU), and low-density polyethylene (LDPE), etc., thermoplastic matrices-based prototypes have been explored for a variety of engineering and biomedical applications. Such thermoplastics and their composite matrix have also indicated good dielectric properties for patch antenna applications. The 3D-printed substrates of PVDF, PLA, PET, LDPE, and ABS (prepared using FDM-based material extrusion and gel 3D printing) have been used to induce a precise band gap in the conducting and ground planes of the antennas prepared. The results obtained from rheological, thermal, mechanical, and morphological investigations of such thermoplastic composites (such as PVDF-6% $CaCO_3$ and PVDF-6% graphene-3% Mn-doped

ZnO) have demonstrated good sensing capabilities for repaired heritage structures. The tunable dielectric properties of these thermoplastic composites are highly acceptable for the manufacture of a variety of piezo-responsive smart sensors. The 3D-printed substrates for the proposed functionally graded composite matrices, prepared by 3D gel printing, laser-based 3D vat polymerization, 3D DIW, extrusion-based fused filament fabrication, etc., may be used effectively for the design and development of conformal piezoelectric sensors. Based on these studies, Figure 12.13 shows schematics for an integrated 3D-printable composite-based piezoelectric sensor that can be tuned for customizable biomedical and engineering applications. The 3D-printed piezoelectric substrate integrates and superimposes the signals from the ground plane and the conducting plane of the sensor to determine the resonant frequency.

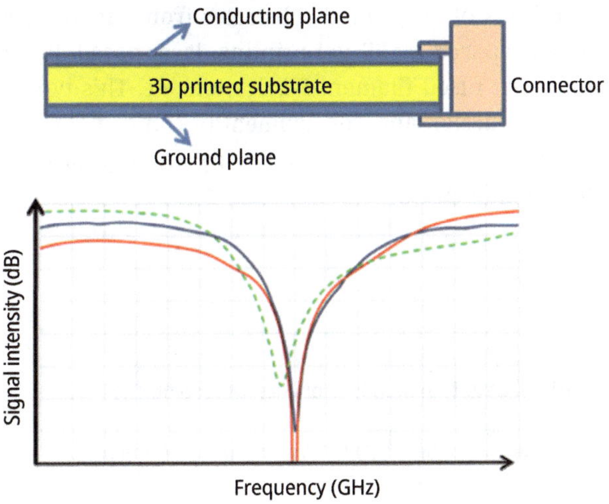

Figure 12.13: Schematics for the integrated 3D-printed piezoelectric sensor and output obtained from the sensor.

The R_f of the fabricated sensor, as shown in Figure 12.13, may be recorded using a vector network analyzer setup. It has been observed that changes in the substrate's dielectric properties may correspond to changes in R_f. Therefore, the manufactured sensor may be tuned for product customization and conformal sensor designing to monitor a wide variety of engineering systems, civil structures, and energy harvesters. Hence, it may be stated that fully 3D-printed and integrated piezoelectric sensors are highly useful for advanced online health/condition monitoring and digital twinning-based industrial and biomedical applications.

12.4 Summary

The studies reported on the design and fabrication of fully and partially integrated 3D-printed piezoelectric sensors highlighted the scope of such sensors in advanced engineering applications. The advantages of piezoelectric pressure sensors, actuators, and tactile sensors in enabling smooth, automated vehicle mobility, consumer product manufacturing, and biosensing clearly demonstrate the wide scope of 3D-printed smart piezoelectric sensors. The approach of fully 3D-printed piezoelectric sensors, which leverages the full integration of multimaterial-based sensor fabrication, is a novel way to manufacture the entire sensor in a single run. This time-saving, cost-effective process enables engineers to design complex sensor geometries alongside the feature to prepare conformal sensors. The integrated approach to 3D-printed piezoelectric sensors outlined the usefulness of 3D-printed substrates from functionally graded polymer composite matrices prepared by 3D gel printing, laser-based 3D vat polymerization, 3D DIW, extrusion-based fused filament fabrication, etc. This type of solution may be used effectively for the design and development of conformal piezoelectric sensors for advanced online health/condition monitoring and digital twinning-based industrial and biomedical applications.

References

[1] Katzir S. The beginnings of piezoelectricity: a study in mundane physics. Dordrecht, The Netherlands: Springer. 2006 Nov 6.

[2] Ahn AC, Grodzinsky AJ. Relevance of collagen piezoelectricity to "Wolff's Law": a critical review. Medical Engineering & Physics. 2009 Sep 1;31(7):733–41.

[3] Dai S, Gharbi M, Sharma P, Park HS. Surface piezoelectricity: size effects in nanostructures and the emergence of piezoelectricity in non-piezoelectric materials. Journal of Applied Physics. 2011 Nov 15;110(10).

[4] Eddiai A, Meddad M, Rguiti M et al. Design and construction of a multifunction piezoelectric transformer. Journal of the Australian Ceramic Society. 2019;**55**:19–24.

[5] Antipov MV, Yurtov IV, Utenkov AA et al. Peculiarities of Performance of Piezoelectric Sensors under a Linear Increase in Pressure. Journal of Experimental and Theoretical Physics. 2020;130:517–22.

[6] Ballas RG. Piezoelectricity. In Piezoelectric Bending Transducers 2025. Berlin, Heidelberg: Springer. https://doi.org/10.1007/978-3-662-70689-3_2.

[7] Hamza M, Kanwal Q, Hussain MI, Khan K, Asghar A, Liu Z, Liu C, Chen Z. Recent progress in 3D printed piezoelectric materials for biomedical applications. Materials Science and Engineering: R: Reports. 2025 Jun 1;164:100962.

[8] Salim M, Salim D, Chandran D, Aljibori HS, Kherbeet AS. Review of nano piezoelectric devices in biomedicine applications. Journal of Intelligent Material Systems and Structures. 2018 Jun;29 (10):2105–21.

[9] Manbachi A, Cobbold RS. Development and application of piezoelectric materials for ultrasound generation and detection. Ultrasound. 2011 Nov;19(4):187–96.

[10] Guo Z. Boundary element method for relative displacement of Eshelby inclusions in piezoelectric materials fused with DEM. Journal of Computational Methods in Sciences and Engineering. 2025;0 (0). doi: 10.1177/14727978251367195.

[11] Wei H, Wang H, Xia Y, Cui D, Shi Y, Dong M, Liu C, Ding T, Zhang J, Ma Y, Wang N. An overview of lead-free piezoelectric materials and devices. Journal of Materials Chemistry C. 2018;6(46):12446–67.

[12] Li S, Shan Y, Chen J, Chen X, Shi Z, Zhao L, He R, Li Y. 3D printing and biomedical applications of piezoelectric composites: A critical review. Advanced Materials Technologies. 2025 Mar;10 (5):2401160.

[13] Smith M, Kar-Narayan S. Piezoelectric polymers: theory, challenges and opportunities. International Materials Reviews. 2022;67(1):65–88.

[14] Das B, Mohanty S. A comprehensive review on piezoelectric inks: From concept to application. Sensors and Actuators A: Physical. 2024 Feb 1;366:114939.

[15] Kumar V, Singh R, Ahuja IP, Hashmi MJ. On technological solutions for repair and rehabilitation of heritage sites: a review. Advances in materials and processing technologies. 2020 Jan 2;6(1):146–66.

[16] He Q, Briscoe J. Piezoelectric energy harvester technologies: synthesis, mechanisms, and multifunctional applications. ACS Applied Materials & Interfaces. 2024 May 13;16(23):29491–520.

[17] Meng Y, Chen G, Huang M. Piezoelectric materials: Properties, advancements, and design strategies for high-temperature applications. Nanomaterials. 2022 Jan;12(7):1171.

[18] Pandey RK, Dutta J, Brahma S, Rao B, Liu CP. Review on ZnO-based piezotronics and piezoelectric nano-generators: aspects of piezopotential and screening effect. Journal of Physics: Materials. 2021 Aug 6;4(4):044011.

[19] Chang G, Pan X, Hao Y, Du W, Wang S, Zhou Y, Yang J, He Y. PVDF/ZnO piezoelectric nano-fibers designed for monitoring of internal micro-pressure. RSC advances. 2024;14(17):11775–83.

[20] Kim M, Fan J. Piezoelectric properties of three types of PVDF and ZnO nanofibrous composites. Advanced Fiber Materials. 2021 Jun;3(3):160–71.

[21] Gubinyi Z, Batur C, Sayir A, Dynys F. Electrical properties of PZT piezoelectric ceramic at high temperatures. Journal of Electroceramics. 2008 Apr;20(2):95–105.

[22] Kumar V, Singh R, Ahuja IS. Use of Thermosetting Polymers for Smart Civil Structures. Encyclopedia of Materials: Plastics and Polymers. 2022a;1:662–69. https://doi.org/10.1016/B978-0-12-820352-1.00067-5.

[23] Ma M, Chen M, Zhang T Boosting the piezo-photocatalytic degradation activity for organic pollutants with SiO2-modified BiVO4 piezo-photocatalyst. Applied Surface Science. 2025 Oct 2:164723.

[24] Mishra P, Kumar P. Effect of sintering temperature on dielectric, piezoelectric and ferroelectric properties of BZT–BCT 50/50 ceramics. Journal of Alloys and Compounds. 2012 Dec 25;545:210–15.

[25] Pala S, Lin L Fully transparent piezoelectric ultrasonic transducer with 3D printed substrate. In 2019, 20th International Conference on Solid-State Sensors, Actuators and Microsystems &Eurosensors XXXIII (Transducers &Eurosensors XXXIII) 2019 Jun 23 (pp. 234–37). IEEE.

[26] Tao R, Granier F, Therriault D. Multi-material freeform 3D printing of flexible piezoelectric composite sensors using a supporting fluid. Additive Manufacturing. 2022 Dec 1;60:103243.

[27] Aliqué M, Simão CD, Murillo G, Moya A. Fully-printed piezoelectric devices for flexible electronics applications. Advanced Materials Technologies. 2021 Mar;6(3):2001020.

[28] Nassar H, Khandelwal G, Chirila R, Karagiorgis X, Ginesi RE, Dahiya AS, Dahiya R. Fully 3D printed piezoelectric pressure sensor for dynamic tactile sensing. Additive Manufacturing. 2023 Jun 5;71:103601.

[29] Bodkhe S, Ermanni P. Challenges in 3D printing of piezoelectric materials. Multifunctional Materials. 2019 Mar 27;2(2):022001.

[30] Kumar V, Singh R, Ahuja IS. Multi-material printing of PVDF composites: A customized solution for maintenance of heritage structures. Proceedings of the Institution of Mechanical Engineers, Part L: Journal of Materials: Design and Applications. 2023 Mar;237(3):554–64.
[31] Liu X, Liu J, He L, Shang Y, Zhang C. 3D printed piezoelectric-regulable cells with customized electromechanical response distribution for intelligent sensing. Advanced Functional Materials. 2022 Jun;32(26):2201274.
[32] Fuh YK, Wang BS, Tsai CY. Self-powered pressure sensor with fully encapsulated 3D printed wavy substrate and highly-aligned piezoelectric fibers array. Scientific Reports. 2017 Jul 28;7(1):6759.
[33] Zeng Y, Jiang L, Sun Y, Yang Y, Quan Y, Wei S, Lu G, Li R, Rong J, Chen Y, Zhou Q. 3D-printing piezoelectric composite with honeycomb structure for ultrasonic devices. Micromachines. 2020 Jul 23;11(8):713.
[34] Martín-Sánchez J, Trotta R, Mariscal A, Serna R, Piredda G, Stroj S, Edlinger J, Schimpf C, Aberl J, Lettner T, Wildmann J. Strain-tuning of the optical properties of semi-conductor nanomaterials by integration onto piezoelectric actuators. Semi-conductor Science and Technology. 2018 Jan 1;33 (1):013001.
[35] Fleming AJ, Leang KK. Integrated strain and force feedback for high-performance control of piezoelectric actuators. Sensors and Actuators A: Physical. 2010 Jun 1;161(1-2):256–65.
[36] Singh R, Barwar A, Kumar R, Kumar V. On mechanically recycled PLA-HAP-CS-based filaments for 3D printing of smart biomedical scaffolds. Journal of the Brazilian Society of Mechanical Sciences and Engineering. 2022a Sep;44(9):416.
[37] Kumar V, Singh R, Ahuja IS. On 3D printed meta-structure-based functional prototype as an innovative solution for repair and online health monitoring of heritage structures. Materials Letters. 2022b Nov 1;326:132950.
[38] Ranjan N, Kumar R, Singh R, Kumar V. On PVC-PP composite matrix for 4D applications: Flowability, mechanical, thermal, and morphological characterizations. Journal of thermoplastic composite materials. 2023 Apr 1;36(4):1401–21.
[39] Mehta A, Singh R, Pabla BS, Kumar V. On 3D printed polyvinylidene fluoride-based smart energy storage devices. Journal of Thermoplastic Composite Materials. 2024 Jun;37(6):1921–37.
[40] Singh R, Singh BP, Singh AP, Kumar V, Kumar R, Bodaghi M, Serjouei A, Wei Y. On 3D printing of low-cost sensors using recycled PET. Sādhanā. 2022b Nov 30;47(4):260.
[41] Mehta A, Singh R, Pabla BS. On 3D Dispenser Printing of Tertiary Recycled PVDF for Sensing Applications. National Academy of Sciences Letters. 2025 Sep;16:1–7.
[42] Singh G, Singh R, Husain M, Singh A. 460 On-Process Capability Analysis of Novel Cerclage Strip for Health Monitoring of Comminuted Fracture in Canines. In In Advanced Manufacturing Technologies in Biomedical Science: Practical Applications, Case Studies, and Future Trends 2026 (pp. 460–67). CRC Press.

Index

 | https://doi.org/10.1515/9783112218846-013